The

Genius

of

Dogs

The
Genius
of
Dogs

Discovering the Unique Intelligence
of Man's Best Friend

Brian Hare

and

Vanessa Woods

ONEWORLD

A ONEWORLD BOOK

First published in Great Britain and the Commonwealth
by Oneworld Publications 2013

First published in the United States by
Dutton, a member of the Penguin Group (USA) Inc.

Hardback ISBN: 978-1-85168-985-9
Ebook ISBN: 978-1-78074-136-9

Designed by Nancy Resnick

Oneworld Publications
10 Bloomsbury Street, London WC1B 3SR

Stay up to date with the latest books,
special offers, and exclusive content from
Oneworld with our monthly newsletter

Sign up on our website
www.oneworld-publications.com

For all dogs

CONTENTS

PART TWO

'DOG SMARTS'

PART THREE

YOUR DOG

PREFACE

When we brought our new baby home from the hospital, our dog Tassie was faced with a dilemma. Since the day we adopted him from a shelter as a puppy, Tassie has had a basket of stuffed toys. Growing up, his favourite activity was to rip out the stuffing and leave it all over the house. Every now and then we would fill up the basket with new toys he could rip up all over again.

We also gave our baby, Malou, a basket of stuffed toys, which was almost identical to Tassie's. As Malou started to crawl, she quickly developed the habit of dragging the toys out of her basket and leaving them all over the house.

Here was the dilemma. Of the dozens of toys, Tassie had to figure out which ones were his to rip up, or Malou was going to find her favourite toys in a heap of stuffing and there would be trouble.

Tassie turned out to be rather good at it. Of course, we were hopeful Tassie would have this ability, since Brian's colleague at the Max Planck Institute for Evolutionary Anthropology in

Germany, Juliane Kaminski, studied a dog named Rico who had solved a similar problem. Kaminski received a phone call one day from a very nice German lady saying she had a Border collie who understood more than two hundred German words, mostly the names of children's toys. This was impressive but not unheard of. Language-trained bonobos, bottlenose dolphins, and African grey parrots have learnt a similar number of names for objects. What was different about Rico was *how* he learned the names of these objects.

If you show a child a red block and a green block, then ask for 'the chromium block, not the red block', most children will give you the green block, despite not knowing that the word *chromium* can refer to a shade of green. The child *inferred* the name of the object.

Kaminski gave Rico a similar test. She placed a new object Rico had never seen before in a different room with seven of his toys that he knew by name. Then she asked him to fetch a toy using a new word he had never heard before, like *Sigfried*. She did this with dozens of new objects and words.

Just like children, Rico inferred that the new words referred to the new toys.

Without any training, Tassie has never ripped up one of Malou's toys instead of his own. His toys and her toys can be lying in a jumble on the floor, and he will carefully extract his toys and play with them, giving her toys only a longing glance or a quick sniff. He adapted quicker than we did to life with a new baby.

In the last ten years, there has been something of a revolution in the study of canine intelligence. We have learnt more about how dogs think in the past decade than we have in the previous century.

This book is about how cognitive science has come to understand the genius of dogs through experimental games using nothing much more high-tech than toys, cups, balls, and anything else lying around the garage. With these modest tools, we have been able to peer into the rich cognitive world of dogs and how they make inferences and flexibly solve new problems.

Thinking about dog genius will not only help us enrich their lives but also broaden how we think about human intelligence. Many of the same concepts used to study dog intelligence are being applied to humans. Perhaps the greatest gift our dogs will give us is a better understanding of ourselves.

Everyone has an opinion about what makes dogs clever. There is now an extensive scientific literature examining dog psychology that sometimes supports or doesn't support these opinions. To help all dog lovers debate what the latest scientific findings might mean, this book provides a comprehensive review of dog cognition, or 'dognition'.

We have read thousands of scientific papers relevant to the study of dog cognition, and we reference more than six hundred of the most important and interesting of these papers in this book. If you are interested, there are ways to get access to these papers and read them for yourself.*

While our review is comprehensive, it covers only areas that have

---

* Most exciting is that much of the research we review is available to you online since: (1) Google has a function called Google Scholar where you can download many of the papers; (2) many scientific journals are more often allowing free access to their papers through their online sites; (3) you can search the website of the scientists who authored a paper to find a link to their publications, where you can download their papers for free; and (4) there is nothing that scientists like more than if you write to them and ask them for their papers. Hopefully, they will happily share if there is a paper we discuss in this book that you cannot get access to otherwise.

been studied scientifically. We may not cover some areas of interest simply because no scientist has published anything on the topic. But on the flip side, there is tons of fascinating research on dognition you may never have imagined.

Although we have done our best to represent the literature fairly, not every scientist will agree with everything we report. Whenever we could, we highlighted alternative perspectives or competing data in the main text. But for ease of reading, we have provided extensive notes at the back of the book that cover important details and alternative findings when they are available.

Disagreement and debate in science are healthy and exciting. Disagreement often drives research that leads to advances in our understanding. Scientists rely on scepticism and empirical debate as a road to the truth. So do not be alarmed if your intuition or your own observations lead you to be sceptical of some of the evidence we present. You are just being a good scientist.

We hope that when you finish this book, your new knowledge, combined with your own observations, will lead to interesting discussions and debates with your fellow dog lovers. Through these debates, we can learn how to have an even richer relationship with our dogs. We can also identify areas where we need more understanding or where scientists have not even asked the right questions. This is all part of the fun.

What we know for certain is that the cognitive world of every dog is far more complex and interesting than we thought possible. We also have a tantalizing glimpse into the secret of their success. We can now pinpoint the stuff of dog genius.

Brian had the good fortune to play a significant role in the unfolding of this story of discovery—as did his childhood dog Oreo.

Some of what is laid out in the following pages will shock even the most knowledgeable dog owners. It is not always obvious where dogs will show an ability to make inferences or show more flexibility than other species. But in the end, your intuition is correct—your dog *is* a genius.

PART ONE

BRIAN'S DOG

1

GENIUS IN DOGS?

The many flavours of genius

Can I really be serious about the title? Most dogs can do little more than sit and stay, and can barely walk on a lead. They are baffled when a squirrel disappears up a tree by circling the trunk, and most will happily drink out of the toilet bowl. This is not the profile of a typical genius. Forget Shakespearean sonnets, spaceflight, or the Internet. If I used the clichéd definition of genius, this would be a very short book.

I am serious, and hundreds of studies and the latest research back me up. This is because in cognitive science, we think about intelligence in animals a little differently. The first thing we look at, when judging the intelligence of animals, is how successfully they have managed to survive and reproduce in as many places as possible. In some species, such as cockroaches, success does not have much to do with intelligence at all. They are just very hardy and excellent reproducers.

But with other animals, surviving takes a little more intellect, and a very specific kind of intellect. For instance, it does not do any good composing sonnets if you are a dodo. You are obviously missing the intelligence you need to survive (in the dodo's case, this was learning to avoid new predators such as hungry sailors).

With this as our starting point, the dog is arguably the most successful mammal on the planet, besides us. Dogs have spread to all corners of the world, including inside our homes, and in some cases onto our beds. While the majority of mammals on the planet have seen a steep decline in their populations as a result of human activity, there have never been more dogs on the planet than today. In the industrialized world, people are having fewer children than ever but are simultaneously providing an increasingly lavish lifestyle for a growing population of pet dogs. Meanwhile, dogs have more jobs than ever. Service dogs assist the mentally or physically disabled, military dogs find bombs, police dogs do guard duty, customs dogs detect illegally imported goods, conservation dogs find scat to help estimate population sizes and movements of endangered animals, bedbug dogs detect when hotels have a problem, cancer dogs detect melanomas or even intestinal cancer, therapy dogs visit retirement homes and hospitals to lift spirits and speed recoveries.

I am fascinated with the kind of intelligence that has allowed dogs to be so successful. Whatever it is – this must be their genius.

What Is Genius?

Most of us have at some time been given a test where scores determine how we are taught or which university we attend. Alfred Binet designed the first standardized intelligence tests in the early twentieth century. His goal was to identify students in France who should receive extra scholastic attention and resources. His original test evolved into the Stanford-Binet Intelligence Scale, which is known as the IQ test.

IQ tests provide a very narrow definition of genius. As you probably remember, IQ tests focus on basic skills such as reading, writing, and analytical ability. The tests are favoured because *on average,* they predict scholastic success. But they do not measure the full capabilities of each person. They do not explain Ralph Lauren, Bill Gates, and Mark Zuckerberg, who all dropped out of university and became billionaires.

Consider Steve Jobs. One biographer said, 'Was he smart? No, not exceptionally. Instead he was a genius'. Jobs dropped out of elite Reed College in Oregon and went to find himself in India; at one point was forced out of Apple, the company he co-founded, when sales were slow in 1985. Few would have predicted the level of his success by his death. 'Think different' became the slogan of a multinational monolith that fused art and technology under his guidance. Jobs may have been average or unexceptional in many domains, but his vision and ability to think differently made him a genius.

Temple Grandin, a professor of animal studies at Colorado State University, is autistic, yet she is also the author of several books, including *Animals Make Us Human.* Grandin has also done more for animal welfare than almost anyone. Although she struggles to read people's emotions and social cues, her extraordinary understanding of animals has allowed her to reduce the stress of millions of farm animals.

A cognitive approach is about celebrating different kinds of intelligence. Genius means that someone can be gifted with one type of cognition while being average or below average in another.

The cognitive revolution changed the way we think about intelligence. It began in the decade that all social revolutions seemed to have happened, the sixties. Rapid advances in computer technology allowed scientists to think differently about the brain and how it solves problems. Instead of the brain being either more or less full of intelligence, like a glass of wine, the brain is more like a computer, where different parts work together. USB ports, keyboards, and modems bring in new information from the environment; a processor helps digest and alter the information into a usable format, while a hard drive stores important information for later use. Neuroscientists realized that, like a computer, many parts of the brain are specialized for solving different types of problems.

Neuroscience and computer technology highlighted the fatal flaws in the idea of a single-dimensional measure of intelligence. People with well-tuned perceptual systems might be gifted athletes or artists; people with less sensitive emotional systems will succeed as fighter pilots or in other high-risk jobs; and those with unusual memories might do well as doctors. This same phenomenon can be observed in mental disorders. There are myriad cognitive abilities that are not necessarily interdependent on one another.

One of the best-studied cognitive abilities is memory. In fact, we usually think of geniuses as people who have an extraordinary memory for facts and figures, since such people often score off the charts in IQ tests. But just as there are different types of intelligence, there are different types of memory. There is memory for events, faces, navigation, things that occurred recently or long ago – the list goes on. If you have a good memory in one of these areas, it does not necessarily mean your other types of memory are equally as good.

For instance, a woman known as AJ (to protect her identity) had a remarkable autobiographical memory. She could remember when and where almost everything happened in her life. When experimenters named various dates, she could report with uncanny precision important personal and public events that occurred – even down to the time of day. But her memory applied only to autobiographical events. She was not a particularly good student and struggled with rote memorization.

In another study, neuroscientists found that London taxi drivers had a higher density of neurons in an area of the brain called the hippocampus. The hippocampus is involved in navigation, and a higher density of neurons means more storage capacity and faster processing. This gives taxi drivers unusual abilities in solving new spatial problems requiring navigation between landmarks.

What makes AJ and taxi drivers worthy of being credited as geniuses is not what standard IQ tests measure. Rather it is their specialized, extraordinary memories.

There are many definitions of intelligence competing for attention in popular culture. But the definition that has guided my research and that applies throughout this book is a very simple one. The genius of dogs – of all animals, for that matter, including humans – has two criteria:

1. A mental skill that is strong *compared with others*, either within your own species or in closely related species.
2. The ability to make *inferences* spontaneously.

Animal Genius – Not All Just Song and Dance

Arctic terns have a genius for navigation. Each year they fly from the Arctic to the Antarctic and back. Every five years a tern will travel the same distance it takes to get to the moon. Whales have an ingenious way of co-operating to catch fish. They create massive walls of bubbles that trap schools of fish, netting them a much heartier dinner than if they hunted alone. Honeybees have evolved a form of dance that allows them to tell other bees where to find nectar-filled flowers – it is certainly a form of genius to be able to make your living by dancing.

Genius is always relative. Certain people are considered geniuses because they are better than others at solving a specific type of problem. In animals, researchers are usually more interested in what a species as a whole is capable of, rather than each individual animal.

Even though animals cannot talk, we can pinpoint their particular genius by giving them puzzles. Animals do not need to talk to solve these puzzles, they just need to make choices. And these choices reveal their cognitive abilities. By presenting the same puzzle to different species, we can identify different types of animal genius.

Since any bird would look like a genius at navigation compared with an earthworm, it helps to compare closely related species. That way, if one species has a special ability that a close relative does not, we can not only identify their genius but also, more interestingly, ask why and how that genius exists.

For example, the spatial memory of Clark's nutcrackers easily rivals the best taxi driver. These birds live at high altitudes in the western US. In the summer, each bird may hide up to 100,000 seeds throughout its territory. In winter, Clark's nutcrackers retrieve the exact same seeds they hid nine months before, even though the seeds are covered in snow.

When compared with their corvid relatives, Clark's nutcrackers are the champions of finding food they have hidden. A tough winter environment has made these birds into geniuses of spatial memory. However, Clark's nutcrackers do not outperform their relatives in every memory game.

Western scrub jays are also part of the corvid family, and they also frequently hide food. Unlike the solitary nutcrackers who rarely steal, however, scrub jays make a habit of it. They watch other birds hide food and later return to steal it. When tested for their ability to remember where other birds had hidden food, scrub jays proved themselves masters while nutcrackers were hopeless in the same situation. Competition has made scrub jays into geniuses of social memory. (Scrub jays do not just pilfer, they also defend against prying eyes. They prefer to hide their food in private, will re-cache their food later in a new location if another bird observes them hiding their food, and even prefer to hide food in darker locations to avoid others seeing them cache it.)

By giving different types of memory puzzles to these closely related species, scientists have been able to discern each species's unique form of genius. By observing the problems each species encounters in the wild, scientists have also been able to understand why the two show different types of genius.

As with people, just because a species looks like a genius in one area does not mean its members are geniuses in other areas. For instance, some ant species are impressive in how they co-operate. Army ants can form living bridges over water, allowing others to cross over on their backs. Other ant species fight wars to protect workers and breeders, and some even 'enslave' other ants, or keep other insects as 'pets'.

But ants have one severe limitation. They are not always very flexible. Most ants are programmed to follow the scent trails of the ants ahead of them. In the tropics, you can find an 'ant mill' where hundreds of thousands of ants walk in a perfect circle that resembles a crawling black hole. Ant mills have been observed up to 350 metres in diameter, with a single lap taking up to two and a half hours to complete. These ant mills are also known as ant death spirals, because often the ants mindlessly follow one another in tightening circles until they exhaust themselves and die. They loyally follow the pheromones of the ants ahead of them to their death.

This leads in to the second definition of genius – the ability to make inferences. Sherlock Holmes was a genius (albeit a fictional one) because even if the solution to a mystery was not clearly apparent, he was always able to find it by making a series of inferences.

Humans make inferences constantly. Imagine speeding towards a crossroads. Even without seeing the traffic light, you can infer the light is red when you see cars entering the road from the other street.

Nature is far less predictable than traffic. Animals have to deal with unexpected surprises. For ants, following the scent of a pheromone trail is usually a foolproof method. But when the pheromone trail becomes circular, ants do not have the mental abilities to realize the trail they are following is going nowhere.

When an animal encounters a problem in the wild, they do not always have time to slowly figure out a solution through trial and error. One mistake can mean death. Hence animals need to make inferences – fast. Even when animals cannot *see* the correct solution, they can *imagine* different solutions and choose among them. This leads to a lot of flexibility. They might solve a new version of a problem they have seen before, or they might even spontaneously solve new problems they have never encountered.

Yoyo is a chimpanzee living at Ngamba Island Chimpanzee Sanctuary in Uganda. She once watched as an experimenter put a peanut through the opening of a long transparent tube. The peanut bounced when it hit the bottom. Yoyo's fingers were too short to reach the peanut, there were no sticks around to use as a tool to reach it, and the tube was fixed and could not be turned upside down. Undaunted, Yoyo made an inference. She collected water in her mouth from the drinking fountain and spit it into the tube. The peanut floated to the top, and she happily gobbled it up. Yoyo realized she could make the peanut float, even though no water was visible when she thought of her solution. In the wild, her ability to make an inference like this could mean the difference between a good meal and starvation.

John Pilley, a retired psychology professor, adopted a Border collie named Chaser. Chaser was eight weeks old and typical of Border collies – she loved to chase and herd, she had intense visual concentration, she enjoyed being petted and praised, and she had limitless energy. Pilley had read of Rico the Border collie who knew at least two hundred German words, previously studied by Juliane Kaminski, and he was interested in seeing if there was a limit to the number of names a dog could learn. Or perhaps the memory of some of the older objects would fade as Chaser learned the names of new objects.

Chaser learned the names of one or two toys a day. Pilley, known as 'Pop', would hold up the toy and say, 'Chaser, this is . . . Pop hide. Chaser find . . .' Pilley did not use food to motivate Chaser. Instead, he used praise, hugs, and play as rewards for finding the right toy.

Over three years, Chaser learned the names of more than 800 plush toys, 116 balls, 26 Frisbees, and more than 100 plastic objects. There were no duplicates, and all of the objects differed in size, weight, texture, design, and material. In total, Chaser learned the names of more than 1,000 objects. She was tested every day, and just to be sure she was not 'cheating' by getting hints from anyone, every month she had to complete a blind test in which she fetched objects in a different room, out of sight of Pilley and her trainers.

Even after Chaser had learnt more than 1,000 words, there was no decrease in the rate at which she learned new ones. More impressive still, the objects were organized in a variety of categories in her mind. Although the objects came in different shapes and sizes, Chaser, without any training, could distinguish between objects that were her toys and those that were non-toys.

We will discuss these studies in greater detail in Chapter 6, but for now, suffice it to say that Rico and Chaser seemed to be learning words in a way similar to human infants – they were inferring that a new word belongs to a new toy. Rico and Chaser knew the new word could not refer to their familiar toys, since they already had names. This left a toy without a name as the only possible answer.

This process of making inferences is critical in understanding how dogs think. In an experimental game, dogs were shown two cups. One of them hid a toy, and the dogs were given one chance to find it. When the experimenter showed the cup where the toy was not hidden, some dogs spontaneously inferred the toy must be in the other cup. In the right situation, many dogs can make this kind of inference.

First, we look for genius in animals by comparing one species to another. Often, the challenges different species face in the wild have provided them with different kinds of genius. Some dance, some navigate, and some have figured out how to have diplomatic relations with other species. Second, we look for genius in animals by testing their flexibility to solve new problems by making inferences.

Genius in Dogs – The Breakthrough

Until recently, science had not taken the genius of dogs very seriously. The abilities of dogs like Chaser and Rico to learn new words could have been discovered as early as 1928. In that year, C.J. Warden and L.H. Warner reported on a German shepherd named Fellow. Fellow was something of a cinematic star; his most memorable scene was saving a child from drowning in the film *Chief of the Pack*.

Much like Rico's owner, who got in contact with my colleague Juliane Kaminski, Fellow's owner contacted the scientists and reported that Fellow had learned almost four hundred words, noting that Fellow 'understands these words in much the same manner as a child under the same circumstances would'. He had raised Fellow almost from birth and talked to Fellow the way you would to a child.

Warden and Warner went to examine the dog. They had his owner give commands to Fellow from the lavatory, so he would not unwittingly give Fellow any extra unconscious cues. They found that Fellow knew at least sixty-eight commands (some of them helpful to a canine film star), such as 'speak', 'stand close to the lady', 'take a walk around the room'. Others were more impressive, such as 'go into the other room and get my gloves'.

The scientists concluded that although Fellow had nowhere near the abilities of a child, more research was needed to understand this type of intelligence in dogs. Unfortunately, their call was not answered until Juliane Kaminski undertook her research on Rico in 2004.

In the intervening seventy-five years, dogs were largely ignored.

When scientists began studying animal cognition in the 1970s, they were more interested in our primate relatives. Eventually, enthusiasm extended to other animals, from dolphins to crows. Dogs were mostly left out of the equation because they were domesticated. Domesticated animals were seen as artificial products of human breeding. Domestication supposedly dulled an animal's intelligence because they had lost the skills and intelligence needed to survive in the wild. Only two research projects were conducted to evaluate dog intelligence between 1950 and 1995, and both found dogs to be unremarkable.

Then in 1995, I did an experiment with my dog in my parents' garage and started something new. I discovered that instead of domestication making our best friends stupid, our relationship with dogs gave them an extraordinary kind of intelligence. Almost simultaneously on the other side of the world, Ádám Miklósi conducted a similar study and independently came to the same conclusion.

These studies caused an explosion in the field of dog cognition. Suddenly, people from a range of disciplines realized what had been under our noses the whole time – dogs are one of the most important species we can study, not because they have become soft and complacent compared with their wild cousins, but because they were clever enough to come in from the cold and become part of the family.

Perhaps the biggest biological mystery of all is the origin of our unlikely relationship with dogs. Every human culture on every continent for thousands of years has included dogs, from dingoes in Australia to basenjis in Africa. Our new understanding of dog genius has provided answers for some big questions about our best friends. How, when, and why did this powerful relationship begin?

And what does it mean when we think about the origins of our own species? And just as importantly, what does it mean for your relationship with your dog?

For the first time, we can hope to answer these questions. To begin our journey, and to understand how this relationship came to exist at all, we must travel back millions of years to a time long before dogs existed. Before wolves and humans had even met.

2

THE WOLF EVENT

Wolves conquer the world, only to lose it all

The best archaeological and genetic evidence suggests that dogs began evolving from wolves somewhere between 12,000 and 40,000 years ago. We take our relationship with dogs for granted, although on closer inspection, it is an astounding thing to have happened. People have proposed that our ancestors adopted wolf puppies, which then became domestic dogs over time. Or that wolves and humans began to hunt together. But neither of these theories really makes sense.

Wolves and humans have never had a particularly amicable relationship, although the animosity is overwhelmingly one-sided. Occasionally, there is a story of children who were adopted and raised by wolves with a happy ending, like Romulus and Remus, who went on to found Rome, or Mowgli in Rudyard Kipling's *Jungle Book*. But for the most part, no other animal has been portrayed so ubiquitously as the Bad Guy throughout history.

The Bible portrayed the wolf as a ravenous destroyer of innocence. In Icelandic mythology, two wolves swallowed the moon and the sun. The Old German word for wolf, *warg*, also means 'murderer', 'strangler', and 'evil spirit'. Those officially pronounced *wargs* were cast out from society and forced to live in the wilderness. Some think this is where the myth of werewolves came from, since the outcast was no longer thought to be human. As children, we grew up with Little Red Riding Hood and the Three Little Pigs, where wolves were cunning villains to be outwitted and slain.

The revilement of wolves was not limited to myths and fables. Almost every human culture in the world that has come into contact with wolves has persecuted them at one time or another, and these persecutions have often led to their local annihilation.

The first written record of a wolf extermination was in the sixth century BCE, when lawmaker and poet Solon of Athens offered a bounty for every wolf killed. This was the beginning of a long, systematic massacre that turned the wolf from one of the most successful and widespread predators in the world to one listed by the International Union for Conservation of Nature (IUCN) as vulnerable to extinction in 1982. (The grey wolf's status had been upgraded by 2004 to a species of 'least concern'.)

The Japanese used to worship wolves and prayed to them to protect their crops from wild boar and deer. When in 1868 Japan ended three centuries of self-imposed isolation from the rest of the world, Westerners arrived and advised the Japanese to poison all of the wolves to protect their livestock. In 1905, three men brought a wolf carcass to sell to an American collector of exotic animals. The wolf had been killed while chasing deer near a log pile. The collector paid the men for the wolf, skinned it, and sent the pelt to London. This was the last wolf in Japan.

Since then, the wolf has been hunted from 80 percent of China and India, and numbers have been dramatically reduced in Mongolia.

In England, the last wolf was killed in the sixteenth century under the order of Henry VII. In Scotland, the forested landscape made wolves more difficult to kill. In response, the Scots burned the forests. Emperor Charlemagne of France organized an order of knights called the *louveterie*, who were essentially wolf hunters. The last wolf in France was seen in 1934.

Wolves fared little better in North America. Wolves were revered and respected in some indigenous tribes, but this reverence did not protect them from being hunted and trapped for fur. The first European settlers brought their prejudices with them, and the war on wolves was swift and thorough. Soon after the first livestock arrived in Virginia in 1609, a bounty was put on wolves. Barely a century later, thanks to traps, strychnine poisoning, and the fur trade, wolves were gone from New England.

In 1915, the eradication of wolves became US government business, and officials were appointed whose sole purpose was the elimination of wolves on the continent. They did their job well. By the 1930s, there was not a wolf left in the forty-eight contiguous states of America. Wolves have been reintroduced recently into Yellowstone National Park and the state of Idaho, though residents in the surrounding communities have successfully lobbied to have them hunted, since wolves occasionally kill livestock.

If this is a snapshot of our behaviour towards wolves over the centuries, it presents a perplexing problem – how was this feared and hated creature tolerated by humans long enough to evolve into the domestic dog?

Domestication requires genetic change over many generations, and the early progenitors of dogs looked very much like wolves – the same animals humans have hunted and persecuted throughout the centuries. When did humans and wolves meet for the first time?

And what happened to convince humans that an animal we traditionally feared and despised would make a good pet? To answer these questions, we have to go back to the very beginning.

Life in a Freezer

Around six million years ago, the earth began to cool. Ice sheets formed on Antarctica and Greenland, and began to build up on North America and northern Europe. In East Africa, some forest-dwelling primates left the trees for more open grasslands. They began evolving to walk upright, which caused myriad changes to their anatomy. They tamed fire, hunted and were hunted in turn, and over millions of years became the face you see in the mirror.

At the same time as our ancestors were coming down from the trees, the first canids appear in the fossil record on the other side of the world in North America. *Canis ferox* was about the size of a small coyote, with a more robust build and a large head.

Aliens visiting the planet six million years ago who happened to observe these two different creatures from their spaceship would not have guessed how closely their paths would intertwine. If you had to choose two species who would share a bond rivalling any other on Earth, you might choose two who shared a longer evolutionary history, who were similar in size, had similar anatomy, even came from the same place. Would you have made the connection between our ancestors standing unsteadily on two feet in the cradle of Africa and a small fanged carnivore on the other side of the world? It was definitely a long shot.

Then, 2.5 million years ago, increasing ice sheets, movement of the tectonic plates, and slight changes in the Earth's orbit around the sun caused the Ice Age. In less than 200,000 years, the world's climate plunged from warm and temperate to freezing. Massive ice sheets as high as 1.9 kilometres covered North America, then collapsed into the ocean, only to form again. Gigantic icebergs filled the North Atlantic, causing temperatures to plunge even further. The formation of the land bridge between the South and North American continents cut off the Atlantic Ocean from the Pacific, and the Arctic and Antarctic waters kept the Atlantic cool and isolated from warmer equatorial currents.

The harsh conditions were interspersed with warm periods, called interglacials, when the climate was much like it is today. A glacial–interglacial cycle lasted around 40,000 years, with variations in between. But at its worst, the Ice Age was not an easy time to live. Forests were destroyed under the ice. The ground was completely frozen, except for a metre or so at the surface that melted every summer then cracked and refroze in the winter. Half the vegetation disappeared. Glaciers cut huge pathways, transforming the landscape and diverting rivers. In addition to the bitter cold, the air was dry and full of dust. Animals and plants retreated from advancing ice sheets towards the equator; then, during interglacial periods, they returned to re-occupy their former habitats.

Between 1.7 and 1.9 million years ago, in this inhospitable environment, a wolf arrived on the scene. *Canis etruscus*, or the Etruscan wolf, was probably the ancestor to modern wolves. Previously, canids were isolated in North America, but the uplift of tectonic plates revealed the Beringian land bridge connecting North America to Asia, and canids quickly crossed into Asia, then spread to Europe and Africa.

The Etruscan wolf was smaller than modern wolves, with a slender build and a skull similar to the American coyote. What is impressive is that this relatively small wolf spread throughout Europe in a conquest so complete, it became known as the Wolf Event.

There were other predators at the time. *Pachycrocuta brevirostris* was the largest hyena ever recorded on Earth. The size of a modern lion, this hyena was the only predator at the time capable of crunching bone with its massive cheek teeth in a powerful skull. But at a quarter of the size of the giant hyena, the Etruscan wolf not only competed with the hyena but went on to become the most successful predator of its time, foreshadowing the success of its descendants.

While the wolves were conquering Europe, early humans were leaving Africa for the first time. *Homo erectus* had large brains and fast-moving limbs and were just beginning to make complex tools. Standing around 1.8 metres, they were a good sixty centimetres taller than their australopithecine ancestors, and their long, gangly legs carried them over the Levantine corridor into Eurasia.

At the archaeological site of Dmanisi in Georgia, beneath the ruins of a medieval fortress, palaeontologists found the remains of our ancestors *Homo erectus*. They also found an almost perfect skull of the Etruscan wolf. Which means it was probably around this time, 1.75 million years ago, that humans and wolves met for the first time.

One million years ago, the Ice Age intensified. Temperatures were erratic, and our ancestors could witness climate change from temperate to freezing within a lifetime. At the more extreme end of the cold period, a giant ice sheet stretched five million square miles, covering the northern part of America from the Atlantic to the Pacific Oceans, reaching all the way down to modern New York City. More ice sheets covered much of northern Europe, extending from

Norway to Russia and from Siberia to northeast Asia. In the Southern Hemisphere, ice covered Patagonia, South Africa, southern Australia, New Zealand, and of course Antarctica.

Cat Kingdom

It was on these giant ice sheets, beneath the shadow of the glaciers, that the Ice Age beasts evolved. Mammals tend to get bigger when it gets cold. Larger animals have a lower surface-area-to-volume ratio, so they radiate less body heat than smaller animals and can stay warmer in colder climates. The second reason mammals get bigger when the climate gets cold is that as the earth cools, it gets drier. Water gets locked up in the ice sheets, and the air cannot hold as much water. This kind of climate is ideal for grasslands. But as the rainfall goes down and the grasslands get drier, the quality of the grass also decreases. Larger herbivores have larger guts that can process low-quality food; they can also roam farther, mowing down vast amounts of vegetation. As an example, a woolly mammoth could spend twenty hours a day grazing up to 180 kilograms of grass.

As the herbivores get bigger, it takes bigger carnivores to bring them down. In Europe half a million years ago, you would have recognized some of the carnivore guild, although you would have been surprised to see them in Europe and probably would have been shocked by their size. The lion (*Panthera leo*) is the same species as the African lion, but 50 percent larger. The hyena (*Crocuta crocuta*) was about 25 percent larger than modern hyenas. Weighing about five hundred kilograms, the cave bear was the largest bear around and was completely herbivorous, although it competed for den space with predators.

Some members of the carnivore guild stayed about the same size. The leopard, *Panthera pardus,* was about the same size it is today in Africa, while wolves were about the same size as the larger modern-day Alaskan wolves.

Then there were species you would not see today. The saber-toothed cat (*Smilodon fatalis*) was the size of modern lions. From the sheer numbers in fossil deposits in the Californian Rancho La Brea Tar Pits (they were five times more common than the next predator down), the saber-toothed cats were the top predators at the time. They gripped their prey with powerful front limbs, using retractable claws to drag the prey towards them. Their upper canines were long and curved and could puncture the neck of their prey with a single lethal stab. They hunted in prides and were capable of bringing down prey much larger than themselves.

Neanderthals – Wild Dogs of the Ice Age

Another member of the carnivore guild was the Neanderthal. These were the Europeans who evolved from the first immigration of early humans from Africa. Their forebears appeared in Europe around 800,000 years ago, and the peak of the Neanderthals began 127,000 years ago. These big, barrel-chested humans had short forelimbs and robust fingers and toes to help them conserve heat and avoid frostbite. They had a rugby ball–shaped head with a heavy browridge and a large lower jaw with a receding chin, giving them an ape-like appearance. Their large, flat noses with massive nostrils probably had an excellent sense of smell and may have warmed the chilling air of the Ice Age before it reached their lungs. They had strong muscled bodies built for carrying heavy loads, but the alignment of their hips shows they probably walked less efficiently than modern humans.

Neanderthals survived the cruellest years of the Ice Age. They hunted woolly mammoths and other large herbivores, and their stone tools allowed them to quickly tear off flesh (similar to wild dogs) and – if they had enough time before the larger scavengers arrived – crack open bones for nutritious marrow, like hyenas.

So this was the Ice Age bestiary. It must have been an awesome sight, the herds of woolly mammoths grazing the tundra, saber-toothed cats lying in wait, and giant hyenas scavenging over a kill. These enormous creatures must have seemed timeless, invincible even.

Then the arrival of a new predator changed everything. Modern humans arrived in Europe around 43,000 years ago, and within 15,000 years, Neanderthals and almost all of the large carnivores went extinct.

There is a lot of argument about what caused this mass extinction at the end of the Pleistocene, particularly of the Neanderthals. Humans have always outcompeted other animals, but it is curious that they drove a close relative to extinction. Far from being the thuggish brutes portrayed in Hollywood films, Neanderthals had even bigger brains than modern humans. They had culture, perhaps language. Although new genetic evidence suggests most European descendants have Neanderthal genes in them, pointing to crossbreeding at one time or another, the larger Neanderthal population definitely went extinct.

Some say it was climate change. Others say it was direct or indirect competition with humans. Steve Churchill, from Duke University, argues that Neanderthals were vulnerable to extinction before modern humans arrived. First, their populations were already thinning in Europe. Their big, squat bodies were good for keeping warm, but they required a lot of calories to maintain, which did not leave much left over for investing in reproduction and child care. Most Neanderthals died between their twentieth and thirtieth birthdays, and Neanderthal bones reveal diseases attributable to malnutrition, such as rickets and osteoarthritis. Thomas Berger, formerly at the University of New Mexico, found that bone traumas in Neanderthals and modern-day rodeo riders were similar, especially in the head and neck areas. Although Neanderthals did not ride bucking horses, they did come into a lot of unfriendly contact with large mammals.

Second, according to Churchill, Neanderthals' diet was mostly meat, which means they were competing with other predators in the carnivore guild, and Neanderthals were not top predators. To be a top predator, you need two things: You have to be big to overpower your competitors, and you have to be social. (For instance, leopards are large but they are not top predators because they are solitary.)

Neanderthals were neither. Though they were robust, they certainly could not compete with lions, saber-toothed cats, or even leopards. And since Neanderthals lived in groups of only fifteen or so, their numbers were not big enough to overpower these predators. Churchill argues that in the predator hierarchy, Neanderthals were probably around the level of a pack of African wild dogs (*Lycaon pictus*) still living on the African savannah today. If they managed to bring down large prey, they quickly had to strip off as much high-quality meat as they could before they were chased off by other predators. Otherwise, they were left scavenging the remnants of other kills.

The consequences of being middle to low ranking are quite

severe. The socially dominant carnivores ate as much as 60 percent of all of the herbivores killed by predators. This means that the rest of the carnivore guild had to split up the other 40 percent. But this split was not even. The next most dominant carnivore got the majority of this 40 percent, then the next dominant species received the biggest share of what remained, and so on. So even though Neanderthals may have been skilled hunters, they still would have struggled to obtain enough meat to survive.

The New Gang in Town

Churchill points out that when modern humans arrived in Europe, they *were* the socially dominant carnivore. Although not able to compete with other carnivores in terms of strength, they came in large numbers. They also had something the Neanderthals did not – projectile weapons, such as spear throwers and perhaps even bows and arrows. Neanderthals had spears, but these spears were close-range weapons. If there was a group of lions or saber-toothed cats on a kill, a small group of Neanderthal males with spears would not have a chance. But large groups of humans who could launch spears from forty to fifty metres away, this was competition.

After modern humans muscled out the carnivores, they fed on the herbivores; woolly mammoths, woolly rhinos, horses, bison, oryx, wild cattle, and red deer. As the modern human population density increased, they started competing for food like fish, birds, rabbits, and squirrels with smaller carnivores, like the lynx and fox, whose numbers plummeted. Later, the large herbivores followed. As a result, 15,000 years after modern humans arrived, most of the larger members of the carnivore guild, including Neanderthals, were extinct.

Only two large carnivores survived – the brown bear and the wolf, *Canis lupus*. The omnivorous brown bear fed on vegetation, fish, and small mammals and perhaps avoided direct competition with humans. Though they did not go extinct, their numbers certainly declined.

The survival of *Canis lupus* is inexplicable. They appear in the fossil record around a million years ago in Alaska, and they arrived in Europe approximately half a million years ago.

Not only did wolves survive to spread throughout most of the Northern Hemisphere and become one of the most successful predators in the world, but somewhere along the line, a subpopulation of wolves spent enough time with humans over many generations that their morphology, physiology, and psychology changed from wild wolves to domesticated dogs.

The theory has long been that humans intentionally adopted wolf puppies and tamed them on purpose. The late zoologist Ian McTaggart-Cowan wrote:

> Somewhere in early history a young wolf was brought into the family circle of man and through the years became the source of the domestic dog and our most successful and useful experiment in domestication.

In a 1974 paper, wolf expert David Mech from the University of Minnesota says,

> Evidently early humans tamed wolves and domesticated them, eventually selectively breeding them and finally

developing the domestic dog (*Canis familiaris*) from them.

But when you think about it, this does not really make sense. Modern humans were extremely successful hunters *without* wolves. And wolves eat a lot of meat, as much as five kilograms per wolf per day. A pack of ten wolves would need an entire deer each day. Starvation was a real threat for many carnivores in the Ice Age, and competition would have been fierce. So fierce that humans, who were no longer content with only 60 percent of the energy budget, drove every other large carnivore except wolves to extinction. (Starvation is a significant cause of mortality in many carnivores, including lions, spotted hyenas, wolves, wild dogs, and lynx. As a general rule, it takes about ten thousand kilograms of prey to support about ninety kilograms of carnivore biomass – irrespective of carnivore body size.)

Wolves are extremely possessive of their food, and if humans wanted some of the kill, they would probably have had to fight them for it. When wolves see prey run, it triggers a 'rush' response, where they chase down their prey and inflict numerous bites until the prey falls. The feeding frenzy that follows is quick and intimidating Wolves are historically plagued by scavengers, and their sharp, shearing teeth are specialized for tearing off large chunks of meat. Wolves prize the same parts of the kill as humans: high-protein internal organs like the liver, heart, and lungs, followed by the muscle. There are frequently squabbles over food, and a bite that would be relatively harmless between wolves could lead to a serious injury if inflicted on a soft-skinned human.

Other domesticated animals make sense. Cattle, pigs, and horses may all have started out wild, perhaps a little aggressive when cornered, but none of them had fangs and lived on meat. The wolf–human relationship makes no sense at all.

And yet, at an ancient site in Israel, to the east of the Mediterranean and north of the Sea of Galilee, a burial site is nestled among the hills beside a lake. Under a slab of limestone, a human skeleton rests its head on its left wrist, and the hand gently lies on another skeleton – a puppy.

Dated at between 10,000 and 12,000 years old, the human was a Natufian, part of a Stone Age settlement along a narrow strip parallel to the Mediterranean Sea that stretched from Turkey to the Sinai Peninsula, on whose tallest peak (Mount Sinai) some say Moses was given the Ten Commandments. It was not the barren, thorny desert it is today, but a forested woodland covered in wild foods and game. The Natufians were hunter-gatherers who settled in the region. They lived in dwellings half buried in the earth and had tools such as knives made from bone and stone grinding implements.

But more important are the burial sites. Tucked away in the heart of the Natufian landscape, each base camp contained graves, either in deserted dwellings or just outside the houses. Bodies were carefully placed, usually stretched out and facing upwards. They might be decorated with headpieces, necklaces, and bracelets made from shells, beads, and teeth. Several graves had more than one body, and the Natufian sites are among the earliest records of humans being buried with another species – in this case a dog. Similarly aged burials of dogs have been uncovered across Europe, the Levant, Siberia, and East Asia.

So somewhere between the arrival of modern humans 43,000 years ago and the first dog burials around 12,000 years ago, wolves became domesticated. Not only that, but the bond between humans and the domesticated wolf – now a dog – was so strong that the two were often buried together. And through the centuries, when wolves continued to be persecuted and almost driven to extinction, dogs and humans grew even closer.

As populations of hunter-gatherers became more sedentary, wolves must have started coming into frequent contact with humans, whether through hunting or scavenging around campsites, or eating rubbish or human faeces. But first, something had to change, something dramatic, so that humans no longer perceived the wolves as a threat.

It was by complete accident that I stumbled on the answer.

3

IN MY PARENTS' GARAGE

The perfect place for a scientific discovery

L ike a Seattle grunge band, I got started in my parents' garage. It was late autumn in Atlanta, Georgia, and we were having a cold snap. The garage had only three walls. The wind cut through my tracksuit and reminded me why garages need a door. Like most garages, ours had a cement floor polka-dotted with oil stains and was full of junk. Cans of paint, toys, and camping gear lined the walls. Dad had strapped an old Mad River canoe to the ceiling so haphazardly, I was convinced it would fall any min ute.

Sitting to one side was my best friend, Orco. My parents got Oreo from a neighbour. As a die-hard Georgia Tech sport fan, Dad liked the idea that our new puppy's Labrador parents were named GT and Jacket. He hoped I would name the puppy Buzz, after Georgia Tech's yellow-jacket mascot. Since I was seven, I named him after my favourite food, Oreo cookies.

There was a fence around our suburban back garden, but it was little more than a symbolic barrier. Oreo could open the gate latch if we forgot to lock it, and there was a corner of the fence low enough for him to jump over. Oreo was always roaming round, getting into trouble. My mother would get the occasional phone call because Oreo had invited himself to a neighbourhood pool party and was splashing about with the kids. Or while driving home, we would find a neighbour trapped behind his lawn mower because Oreo was endlessly plopping tennis balls in front of it to force a game of fetch.

But Oreo only got into trouble when I was not around. He preferred hanging out with me to anything else. We rambled through the woods, swam in neighbourhood lakes, and visited my friends and their dogs. Oreo was so loyal that when I rode my bike to a friend's house for a sleepover, Oreo would sit on the doorstep until we went home the next morning.

I was as obsessed with baseball as Oreo was with me. We were a match made in Cooperstown. I could take a bag of balls, hit each one, and wait till Oreo brought them back so I could do it all again. Or I would take aim at something in the garden, and whether I hit or missed, Oreo brought back the ball so I could keep throwing. I knew at ten that my career as a starting pitcher for the Atlanta Braves would be thanks to Oreo. He never let me quit. If I put away the balls, he would conjure another one, drop it on my feet, and bark until I was ready to go again. The one hitch was that any ball that entered Oreo's mouth became a drool sponge. By about the tenth throw, the ball would be twice its original weight, with a slobber tail like a comet. Oreo probably never understood why no one was as excited to play ball with him as I was.

To Be or Not to Be Human

Fast-forward ten years. I made the first cut of my university baseball team but soon gave it up. A professor at Emory University in Atlanta had introduced me to something that captured my imagination more than winning the World Series. Mike Tomasello was trying to figure out what makes us human. As a nineteen-year-old, I had never given much thought to such a profound question, and I was in awe that anyone could even attempt to answer it.

There is no doubt that our species has a special kind of genius. Our powers are not always used for good, but they are impressive. We have managed to colonize every corner of the Earth, moulding glaciers and deserts into comfortable living quarters. We are the most successful large mammals on the planet in terms of our population size and the influence we have on the environment. Our technology can preserve or destroy life. We can fly above the Earth's atmosphere and trawl the deepest trenches of the ocean. As I write, *Voyager 1* is more than eleven billion miles from our planet, sending NASA signals from the edge of our solar system.

It was not always like this. A few million years ago, we were indistinguishable from other forest-dwelling apes. Fifty thousand years ago, we were dodging the fangs of saber-toothed cats and giant hyenas. Twenty thousand years ago we had no governments or even permanent housing. Today, many of us cannot imagine surviving without the Internet or iPads. What happened to us after our ancestors' split from the last ancestor we shared with other apes? What was the first change that led to all of the other changes? How did it all happen?

Until I met Mike, I had not realized that to understand what it is to be human, you had to figure out what it is *not* to be human. My new vocation was studying the minds of other animals to better understand ourselves. Likewise, as a famous psychologist studying the development of infants, Mike compared infants with chimpanzees to test ideas about what makes us unique. He never guessed that he was destined to become a dog researcher.

It was Oreo who led Mike and me to our destinations, but it was Mike's knowledge of infants that led us to Oreo. The theory and methods for studying infant psychology allowed for the revolution in our understanding of dogs.

Social Networking

Humans are not born with adult cognitive abilities. Our infants are born helpless and require the greatest amount of parental care of any animal. This is mostly due to babies' underdeveloped brains. At birth, our brains are only a quarter of their adult size. This is because the human pelvis is designed for walking upright, which has resulted in a small pelvic aperture relative to bonobos and chimpanzees. The aperture is so small that only underdeveloped brains can fit through it at birth. This means most of our brain growth must occur afterwards.

Studies in cognitive development have revealed that not all skills develop at the same time and rate in infants. Early skills become the foundation for more complex skills.

Mike was one of the first to realize that infants develop powerful social skills as early as nine months old. This nine-month revolution allows infants to escape an egocentric view of the world. Infants begin paying attention to what others are looking at, what others are touching, and how others react to different situations. If a mother looks at a car, her infant begins following her gaze by matching his line of sight to hers. If an infant sees something strange like an electronic singing Santa, before he reacts, he looks to an adult's face to gauge the adult's reaction.

Almost simultaneously, infants begin to understand what adults are trying to communicate when they point. Infants also begin pointing out things to other people. Whether infants watch you point to a bird or point to their favourite toy, they are beginning to build core communication skills. By paying attention to the reactions and gestures of other people, as well as to what other people are paying attention to, infants are beginning to read other people's intentions.

Intention reading provides a cognitive foundation for all human forms of culture and communication. Shortly after the nine-month revolution, infants begin to imitate the behaviour of others and acquire their first words. Intention reading allows infants to accumulate cultural knowledge that would be impossible for them to obtain on their own. Infants who show delayed development of intention reading usually have problems learning language, imitating, and interacting with other people. Without culture and language, we could not build on the accomplishments of previous generations. There would be no laws, rockets, or iPads. We would probably have ended up as easy targets for all sorts of predators.

For example, once when I was standing on a beach in Australia, I saw a large black fin rise out of the water near a group of swimmers. Over the noise of the waves, it was impossible for them to hear me. I waved frantically at the swimmers, and when I had their attention, I did something I had never done before – I bent over and placed my hand on my back like a fin. The swimmers hurriedly got out of the water, even though they had probably never seen anyone do this. From my simple gesture, the swimmers knew I had seen something they had not. Given the context, they were able to infer what I was trying to tell them – there was danger in the form of a shark. This social inference requires an understanding of my *communicative intention*. The swimmers understood my gestures as both communicative and helpful. They could then think about how they should adjust their behaviour. Luckily, the fin belonged to a dolphin, but if it had been a great white shark, understanding communicative intention could have saved their lives.

Understanding communicative intention gives us unprecedented flexibility in solving problems. To find out if this is what makes us human, Mike compared humans with our closest living relatives, bonobos and chimpanzees. If we have a skill that bonobos and chimpanzees do not have, then this skill probably evolved after our lineage split from the bonobo and chimpanzee lineage between five and seven million years ago.

Mike needed to compare the abilities of bonobos and chimpanzees in understanding communicative intentions to infants'. If understanding communicative intentions is as crucial to humans as Mike thought, then bonobos and chimpanzees should not show an understanding of communicative intentions. However, if bonobos and chimpanzees were as skilled as young infants, Mike would know he was on the wrong track.

It is quite tricky to test for an understanding of communicative intention in an ape who has no language. However, while human language is the most complex form of communication, it is not the only form. Both bonobos and chimpanzees use visual gestures during social interactions. They can ask to play by pushing or slapping someone; they can ask for food by placing their hand under the chin of someone who is eating. By the time they are adults, bonobos and chimpanzees use and understand dozens of different gestures. This is similar to infants. By examining how bonobos and chimpanzees respond to the gestures of others, we can see whether they understand one another's intentions.

Mike borrowed a game developed by Jim Anderson, a Scottish primatologist at the University of Stirling. Anderson hid food in one of two containers, then gave various primates a clue where the food was. He either touched, pointed, or looked at the container that had the food. Anderson tested capuchin monkeys, who failed miserably. To succeed, the monkeys had to be trained over hundreds of trials, and each time they got a new kind of clue, they had to be trained all over again.

Since chimpanzees are so socially sophisticated and so closely related to us, Mike thought they would do better than monkeys. But chimpanzees also failed. Even if they eventually learned they should choose the container you pointed at, if you changed the clue by standing farther back from the container while you pointed, again the chimpanzees failed.

The only exception was when the chimpanzees had been raised by humans. This meant that they had interacted with humans for thousands of hours. The few chimpanzees with this unusual rearing history were the only ones who were able to use a variety of human gestures to find the food spontaneously.

Mike seemed to have strong support for his idea that spontaneously understanding another's intended communication was a kind of genius unique to humans. Unlike infants, chimpanzees could use new gestures in a new context only if they were given a great deal of practice with the game or if they had been raised by humans.

This suggested that chimpanzees do not understand that when you point, you are trying to help them. Mike thought he may have discovered what made humans unique.

My Dog Can Do That

One day in my second year at university, I was helping Mike play these signalling games with chimpanzees. We started to discuss the implications of our findings. Mike suggested that only humans understand communicative intentions, which allows us to spontaneously and flexibly use gestures, such as pointing.

I blurted out, 'I think my dog can do it.'

'Sure.' Mike was amused. 'Everybody's dog can do calculus.'

During our 'baseball training', Oreo developed a special talent. He could fit three tennis balls in his mouth at the same time, sometimes four if he positioned them just right. I would throw one, and then while he was fetching the first one, I would throw the second and third in different directions. After he had fetched the first ball, Oreo would look at me, and I would point to where the second ball was. After he fetched the second ball, I would point to the third, and off he would go to find it and finally bring all three back to me in triumph, his cheeks bulging like a chipmunk who had eaten the whole bag of nuts.

This did not seem much different from the game the chimpanzees were failing. Oreo was using my pointing gestures to find the tennis balls.

'No, really. I bet he could pass the tests.'

Seeing I was serious, Mike leaned back in his chair.

'Okay,' he said. 'Why don't you pilot an experiment?'

I took Oreo and a video camera to the neighbourhood pond where we often played fetch. I threw a ball into the middle of the pond. When Oreo got the ball, I pointed to the left. Since I often threw two or three balls, Oreo went in the direction I pointed. Then I pointed right, and again, he used my pointing gesture to look for the other ball. I did this ten times, and each time Oreo went in the direction I pointed.

When Mike saw the video, he called developmental psychologist Philippe Rochat into the room. They watched and rewatched in amazement as Oreo effortlessly did something they had come to believe only humans could do.

Mike could not have been more excited. 'Now let's really do some experiments.'

The Importance of Experiments

Experiments can be viewed as a microscope into other minds. Behaviour that appears to be the same between two individuals or two species can be driven by different types of cognition. As Mike wrote:

> To test the flexibility of some behavioral skill we need to expose individuals to new situations and see if they adapt their skills in flexible and intelligent ways.

Experiments give us a way to choose between competing explanations for an animal's cleverness by presenting the same problem in at least two slightly different ways. Variables are carefully controlled in both situations, except for the factors you wish to investigate.

The first scientists to study animal intelligence at the turn of the twentieth century quickly realized the importance of experiments. Lloyd Morgan, one of the most famous of these scientists, used his dog Tony as an example. One of Tony's talents was opening the gate to Morgan's garden. If you watched Tony open the gate, you might assume Tony was pretty clever and understood how gates worked (that is, if the latch is connected to the fence, the gate cannot move). However, Morgan had seen Tony's long process of trial and error and concluded that Tony did not understand why he was able to open the gate; he just got lucky and found a way to open it by accident.

Without an experiment, choosing between a cognitively richer or simpler explanation of Tony's behaviour becomes a matter of opinion. The scientific method favours the simplest explanation in any field, and Morgan is often given credit for illustrating the power of simple cognitive explanations even when studying seemingly complex behaviour.

My Tony moment came while I was working in Rome with a capuchin monkey named Roberta. Elisabetta Visalberghi, the world's expert on capuchin cognition from the Institute of Cognitive Sciences and Technologies in Rome, had given Roberta a problem to see if capuchin monkeys could make spontaneous inferences while using tools. To solve the problem, Roberta and the other monkeys had to get a peanut out of a transparent tube. At the bottom middle of the tube there was a small trap. To get the peanut out of the tube, the monkeys had to insert a tool into the opening of the tube farthest from the peanut while pushing the peanut away from the trap towards the opposite opening.

Only Roberta learned to solve the problem. It looked like she might be some kind of monkey genius, but being a good experimentalist, Elisabetta did another test. She flipped the tube round so the trap was above the peanut and no longer functional. If Roberta understood that it was the trap that prevented her from getting the peanut, she should no longer worry about which end she inserted her tool into. She could push the peanut in either direction and always succeed. Yet, Roberta continued to use the strategy she developed when the trap was functional. She always poked the stick into the end of the tube farthest from the peanut so that she could push the peanut away from the trap.

The experiment showed that Roberta could solve the problem but

she did not understand why the problem occurred (no shame, since as I type on my computer I have no idea how it works, either).

Even though a simple explanation often underlies complex behaviour, this is not always the case. In fact, Morgan was so horrified at the response to his writings – early psychologists used them to say that animals are incapable of making inferences – that he added to his general principle:

> It should be added, lest the range of the principle be misunderstood, that the canon by no means excludes the interpretation of a particular activity in terms of the higher processes, if we already have independent evidence of the occurrence of these higher processes in the animal under observation.

Morgan would be pleased to learn that while Tony did not understand how the latch worked on the gate, scientists have discovered that dogs do not solve this type of problem only by means of trial and error. A recent experiment has shown that dogs can solve the latch problem immediately if they see someone else solve it first. Tony's case demonstrates how experiments can reveal where animals are geniuses and where they are not.

Because of a long history of experiments with animals like Roberta the capuchin monkey, Mike knew that sometimes when animals look intelligent, a slight twist on the problem shows they really have no understanding of what they are doing. To find out what Oreo really understood, then, we needed to conduct a series of experiments. Just because Oreo's behaviour looked similar to an infant's, it did not mean that Oreo understood the communicative intentions behind the gesture the same way infants do.

From Garage to Revolution

So there we were, Oreo and I in the autumn of 1995, in my parents' freezing garage. We decided to use the same tests with Oreo as researchers used with infants, monkeys, and chimpanzees. I placed two plastic cups about two metres apart, faked putting food under one cup, and surreptitiously placed food under the other one. Then I did something Oreo had never seen before.

I stood in the middle of the two cups and pointed to the one that had food inside. (The girl in the illustration below is a paid professional who looks nothing like I did back then.)

'Okay, Oreo, go find it!'

Oreo went straight to the cup I pointed at. I did it again and again, and Oreo always followed my point. Oreo was solving a new problem spontaneously. It was possible he was making a social inference about the meaning behind my gesture.

'Hey, Oreo, you're a genius!'

Oreo pushed his big, warm body against my legs and slobbered all over my face. His soggy tongue left crumbs of dog biscuits on my cheek. It was a great moment in my scientific career.

I busted into Mike's office and showed him the results. Mike was beside himself with excitement, even though it meant his hypothesis – that only humans understood communicative intentions – might be wrong. But there were still explanations we needed to rule out before we could conclude that Oreo understood communicative intentions.

A million questions came to mind. Did Oreo just smell where to find the food? Did Oreo slowly learn how to follow gestures over time? Like chimpanzees, could he use only one type of gesture, inflexibly? Did he just turn his head in the direction my arm moved and then move in the direction where he was looking? Or was he doing something far richer? Did Oreo understand I was trying to help him? Did Oreo understand my communicative intention to show him the location of the food? Did Oreo know that I knew where the food was, even though he did not?

Autumn turned to winter. Trees turned to skeletons, and the bitter wind swept the dead leaves across our drive. Though the snowless Georgia winter was fairly mild, I wore thermal underwear, flannel trousers, a down coat, and gloves, and I still could not feel my fingers as I arranged the cups in the unheated garage. Oreo, on the other hand, with his thick, black coat, worked even better in the cooler weather.

The first thing we had to do was make sure Oreo did not smell

where to find the food. Again, I hid food in one of the cups, but this time, instead of pointing, I looked down at the ground.

Although Oreo could probably smell food in the vicinity, he could not track it down to the exact cup when he made his first choice. When I did not point, he chose the cup with food first only around half the time. Since the probability of finding the food on his first try by chance was fifty–fifty, we knew he was guessing. Subsequently, more than a dozen studies from seven different research groups that have tested hundreds of dogs have thoroughly ruled out the possibility that dogs use their sense of smell to find the food in this context.

Maybe Oreo was just associating my hand with food and choosing the cup closest to my extended finger. I put three cups in front of him and three cups behind him, and pointed to a cup behind him. Oreo turned around and chose the cup behind, even though he had to go *farther* away from my finger to get to it.

Maybe during his many interactions with me, Oreo had learnt to inflexibly use a few signals. If this was true, he should show improvements during the tests and would have trouble reading cues he had not seen before. But Oreo almost always chose correctly on the first trial and did not show improvement as he went along, because he was nearly perfect from the start. And he had no problem using cues he had not seen before – say, if I pointed with my foot. Or even if I just turned my head and looked at the cup.

Maybe Oreo's response was reflexive, and Oreo was just reacting to the movement I made with my gestures, so if I moved a part of my body to the left or right, it made Oreo shift his gaze in the same direction in response to the motion. If this was true, Oreo would approach the cup he was looking at without understanding my attempt to help him. But Oreo followed my pointing cue even if I pointed to the left while taking a step to the right. This meant that even as he watched me move *away* from the correct cup towards the

incorrect cup, he had to go in the opposite direction to his gaze to follow my point. I even had my little brother Kevin cover Oreo's eyes, so that when he opened them, I was already pointing at the cup – there was no motion, and Oreo could use this static pointing gesture. Subsequent research has shown that dogs are also skilled even if you point towards the correct cup only momentarily but then stop pointing while the dog is choosing where to look. It did not look like Oreo was simply reacting to the motion created by human social gestures.

Mike was impressed, but he thought we needed a dog who did not play fetch as much as Oreo. Perhaps over the years, our ball games had given Oreo a chance to slowly learn to use human gestures flexibly. So I recruited my little brother's dog, Daisy.

Daisy was a cuddly, black, rescued mutt who did not play fetch at all. If you threw the ball for her, there was a chance she might lope after it, but she certainly did not bring it back with any enthusiasm.

Daisy proved to be almost as good as Oreo on all of the tasks. She spontaneously used my gestures and gaze direction to find the food and passed most of the controls Oreo had passed. It looked like Oreo's skills were not unusual and were probably possessed by many dogs. This meant we could recruit a larger group.

We wanted to know if dogs could use only their owner's gestures or if they could also use the gestures of a stranger. Just as people can unconsciously react to the quirks and habits of their dogs, maybe dogs slowly learn to respond to the idiosyncrasies of their owners. This would mean that dogs would not respond to the gestures of other people. To test this idea, I went to doggy day care.

A popular doggy 'day care' centre near Emory University was Our Place or Yours Pet Services. Here people could drop off their dogs to play while they went to work. There were dozens of dogs there, and I was a stranger to all of them. These dogs had not grown up watching my body language the way Oreo and Daisy had, and maybe they would find the cues of a new human harder to read.

Remarkably, this was not the case. The dogs followed my gaze and pointing gestures as skillfully as Oreo and Daisy. Other research groups have since received similar results. Dogs can use the gestures of just about anyone.

The next step was to find out if dogs could use the gestures of other dogs. Maggie was a yellow Labrador at the doggy day care. On account of slight arthritis, she was not exactly a wiggle worm. We realized if we anchored her lead to a wall hook, she would stand perfectly still while facing two cups. While Maggie was watching, we hid food under one of the cups. Maggie looked towards the correct cup, but her body was equidistant between the two. We then let another dog approach one of the two cups. This second dog easily solved the problem and spontaneously followed Maggie's gaze and body orientation to find the food.

Now Mike and I knew we were really on to something. We had ruled out a whole bunch of simple explanations for why Oreo understood gestures better than chimpanzees, we knew his skills were shared with other dogs, and these skills could be applied to interactions with other dogs. It was time to give a group of dogs an even more challenging test.

In testing whether chimpanzees understood communicative intentions, Mike used a new clue. Chimpanzees watched a human place a small block on the correct cup. The block was a new arbitrary signal. The chimpanzees had never seen it before, and it had no particular meaning. To use this new arbitrary gesture, the chimpanzees would have to make a generalization and infer that the new gesture had similar meaning to other gestures they had learned to use. This did not happen: The chimpanzees did not guess what the block meant and did not use this clue to find the food.

We decided to run this study with the day-care centre dogs. I painted a black-and-white cow pattern on a block of wood. It was unlikely that any of the dogs had a similar toy and also unlikely that their owners had used a similar signal. In a control situation, when the dogs entered the room, one of the cups had the block sitting on top. The dogs did not choose the cup with the block more than the other cup, so we knew they were not just attracted to the block. The block was a new signal with no particular meaning. But if the dogs saw a human actually place the block on one of the cups, they spontaneously used this strange signal to find the hidden food.

Then we made it more difficult. Just in case the dogs made use of the arbitrary gesture because they were simply attracted to the motion created by placing the block, a barrier blocked the dogs' view of the human placing the block. After the block was placed, the humans touched the block. The dogs still used this clue to find the food.

To be doubly sure, we repeated the study with a different group of dogs in Germany and compared their performance to a group of

chimpanzees on the same test. The dogs spontaneously preferred the cup with the block that the human had touched, while the chimpanzees did not. Dogs were better at understanding our gestures than chimpanzees.

Are Dogs Really Just Like Children?

Even though dogs were making the same choices as human infants, the goal of our experiments was to see how similar the thought process that led to those choices was to that seen in infants. Infants infer that the human giving them the clues is trying to communicate with them. So far, the evidence from our experiments supported the idea that dogs were doing something similar.

Others have tested this idea further and found that dogs are selective and do not just use any kind of clue. They spontaneously use only communicative gestures and ignore other non-communicative clues. For example, dogs will use your gaze direction if you look directly at the correct cup. In contrast, if you turn your head in the direction of the correct cup, but gaze above it instead of right at it, dogs do not use your gaze to find food.

Dogs are most likely to follow your gaze if you indicate your communicative intentions by calling their name and looking directly at them before indicating where they should look. Dogs are also more likely to use your pointing gesture if you make eye contact with them before gesturing or if you alternate your gaze between them and the correct cup while pointing.

These studies suggest that dogs interpret your gesture depending on what you are paying attention to. In short, Mike and I concluded that dogs have communicative skills that are amazingly similar to those of infants.

The Origins of Dog Genius

I graduated from Emory and began work on my PhD. My new adviser at Harvard was British anthropologist Richard Wrangham. I met him while studying chimpanzees at his field site in Uganda during my undergraduate studies. There was no one exploring dog cognition in the US at the time, but my research sparked Richard's interest. Mike moved to Germany to the Max Planck Institute for Evolutionary Anthropology, and Mike and I agreed that while I was studying under Richard, I would fly to Germany to work with Mike and the four great ape species at the Leipzig zoo. When I was in Boston, I spent my free time trying to understand the origins of the unusual communicative skills we had uncovered in dogs.

The obvious explanation for why dogs are so unusual is their peculiar rearing history. According to our 'exposure hypothesis', while dogs may not have learned our gestures during our test, they slowly learned to use them during the thousands of hours they had spent as part of a human family. Just as chimpanzees who were raised by humans could spontaneously pass the gesture tests, perhaps dogs raised by humans had learnt the same skills. The exposure hypothesis predicts that puppies will slowly improve in their ability to use human gestures as they grow older, and their skills will improve as they spend more time with humans.

To test this hypothesis, each weekend I had to be licked to death by dozens of ridiculously cute puppies. To make sure puppies would be able to observe my gestures, I had to get down on their level. I would either point or gaze at one of the two cups while lying on my stomach. This made me vulnerable to high-energy puppy attacks.

Not only were the puppies cute, their performance shocked us – they were as skilled as adult dogs at spontaneously using gaze and pointing gestures. They did not improve with age. Nine-week-old puppies were as skilled as twenty-four-week-old puppies when using a basic pointing gesture.

The amount of time the puppies spent with humans did not matter, either. When I compared puppies adopted by a human family with those living with their litter, there was no difference in performance. Even though the litter-reared puppies had relatively little exposure to humans, they still were performing nearly perfectly. Research since this original study has shown that puppies as young as six weeks of age can spontaneously use different types of human gestures. This is truly amazing since at this age, puppies have barely opened their eyes and are just learning to walk.

Another study found that it did not matter whether dogs were kept inside or outside of the house, or whether they received explicit training or more daily attention than other dogs. All groups were skilled at understanding a human pointing gesture. Even rescue dogs are just as skilled at using a variety of human social gestures as family-reared dogs.

More remarkable still, puppies as young as six weeks old pass the block test. Just like adult dogs, if puppies saw a human put the block on a cup, this was the cup they preferred. The puppies could also use a pointing gesture if the human was standing a metre away from the cup they were pointing at, or the puppies had to go away from the human to approach the correct cup. This rules out the possibility that puppies were choosing correctly just because they are attracted to human hands.

None of these results was what we had expected to find. Rarely in the study of animal cognition has a skill been observed at such an early age that is largely unaffected by different rearing histories. It would have made sense if the skill we observed in dogs was related to their lifestyles, but there was no evidence to support this idea. We will see later that some breeds can be affected by training and dogs can get better as they get older, but for the most part, puppies are already so good with human gestures that there is little room for improvement.

With no evidence to support our exposure hypothesis, we started to think about other explanations for the origin of the unusual abilities of dogs.

'Wolves,' said Mike over the crackling phone line from Germany. 'Now you need to find wolves.'

Yikes.

Running with the Pack

If you are going to live in a carnivorous pack, it is important to predict what other pack members might do next. Gestures are a type of social information that help us guess what someone else might do. The direction you are looking can predict where you will go, and where you are pointing can indicate the object you are interested in. Knowing what someone will do next helps us co-ordinate our behaviour. If wolves have the same skills as dogs, it could help them co-ordinate during a hunt.

For a carnivore, it would also be useful to predict your prey's next action using *their* social information. If you see a deer look left, you can head in that direction and get there before the deer does. Dogs might be unusually skilled at reading human gestures because they come from a group of carnivores that makes a living reading the social information of other species. According to our 'pack ancestry' hypothesis, as the direct ancestor to dogs, wolves should be as skilled at reading human communicative gestures as dogs.

Mike knew it might be tough for me to find the right wolves. Wolves by nature are extremely wary of humans. Even wolves born in captivity are nervous around humans. Previous work comparing the learning ability of dogs and wolves showed that the results depended on the rearing history of the wolves. Wolves reared by their mothers perform worse than dogs, while wolves reared by humans can outperform dogs on the exact same learning task.

We had already tried to test a pair of mother-reared wolves in Germany. They failed to use our gestures, but they would probably fail any test we gave them since they had no interest in interacting with us. We knew if we were going to compare the cognition of wolves and dogs, we had to find wolves who grew up interacting with people in a way that was similar to dogs.

I scoured the Internet and miraculously stumbled on the perfect pack. Wolf Hollow was started by a firefighter named Paul Soffron, who was given five wolf pups in 1988. He created a wolf sanctuary in Ipswich, Massachusetts, to educate and enchant the public about the many reasons to love wolves.

By the time I arrived, Paul had thirteen wolves and an educational centre that attracted 30,000 visitors a year. He was also bedridden, stricken with Alzheimer's. His wife, Joni, was running the place, and she was excited about the research. I also met Christina Williamson, a young biologist with strawberry-blond hair and blue eyes. Small and slender, Christina was not the type you would think would be running with a wolf pack. But she had raised many of the wolves from pups and knew the rearing history of the others. She was as close as you could be to a pack member without actually living in the group.

The wolves had an unusual amount of exposure to humans. Most importantly, they were reared by humans for the first five weeks of their lives, without their mothers. Then they were integrated into a wolf pack while still interacting with Christina almost every day. I watched in amazement as Christina went into the enclosure to separate some of the wolves so they could play our gesturing game. Unless I was going to raise a group of wolves myself (not ideal for a Boston apartment), this was more than I could have hoped for.

As I sat down in front of my first wolf to show Christina how the test worked, I remember thinking, *What big* everything *you have, Grandma!* I was struck by the cautious way the wolves approached. Even though they saw dozens of new people each day as part of the education programme, they were visibly sizing me up. When I pulled out their favourite treat – cheese squares – their demeanour changed. Now I had two wolves in front of me when I only wanted one. Before Christina could say *Stop!* I tried to give a piece of cheese to each wolf. Teeth flashed as one wolf snapped its jaw over the muzzle of the other wolf, who shrieked. There was no warning growl, just a full-force bite. These were definitely not dogs.

After I showed Christina the method of the test, I asked if she could try it. The wolves approached her immediately and let her scratch them through the fence with their tails wagging. It was clear that Christina would be doing most of the experiments.

When we had the results, I shook my head. I had really thought wolves would do just as well as dogs, maybe better. The pack ancestry idea made sense. Instead, the wolves looked like chimpanzees. Even though Christina tried three different types of gestures to show them which cup hid the food, the wolves chose randomly. The nine-week-old puppies I had tested showed more skill at reading human gestures than the wolves had.

At the time, not many people had studied wolf cognition, so we were worried the wolves were just bad at any test given by a human. However, when Christina played another game, where the wolves had to remember which of her hands hid food, they almost never made a mistake. Their failure to use Christina's gestures was not because they were disinterested in food games or were anxious around humans.

Subsequently, researchers have raised wolves for the sole purpose of comparing their social skills with those of dogs. These researchers gave the wolves even more exposure to humans than the wolves I tested, but their results have been similar to ours. At four months of age, heavily socialized wolves could not use a caretaker's gesture to help them find food, even though the caretaker had raised them from puppies. When the wolves were tested as adults, they needed explicit training to match the spontaneous performance of dog puppies. Like chimpanzees, wolves can learn to use human communicative gestures with training or socialization, but they do not spontaneously demonstrate this skill without any training.

József Topál of the Hungarian Academy of Sciences found another way that dogs' dependence on our social information makes them more like infants than wolves. The unusual skill of dogs causes them to make the same errors that infants make.

Dogs watched an experimenter hide a toy in one of two locations. The dogs easily found the toy. Then the experimenter hid the toy behind one location – but then in full sight of the dog moved the toy to the second location. Just like infants, dogs incorrectly searched in the original hiding spot, even though they had seen the toy move. When the hiding was done by means of transparent strings instead of a human, dogs, like human infants, no longer made this error. The error is caused by the social context, not a lack of memory.

Intriguingly, wolves raised by humans did not make the error that infants and dogs make. They were almost perfect at finding the toy even when the human had moved it. Topál and colleagues concluded that this supports the idea that dogs have evolved an unusual sensitivity to human social information that makes them similar to infants. It also shows how relying too heavily on humans can get dogs confused in some situations. Wolves have their own kind of genius.

In these studies, dogs were not just unusual compared with primates, they were unusual compared with their closest canine relatives. They cannot have simply inherited their unusual infant-like abilities. With no support for the exposure hypothesis or the pack ancestry hypothesis, there was only one explanation left. During the process of domestication, dogs evolved a basic understanding of human communicative intentions.

This was an exciting idea, since it suggested that the cognitive abilities of dogs converge with those observed in infants. *Convergence* is when distantly related species independently evolve similar solutions to the same problem. Biologists have often found convergence in physical structures of distantly related species. For instance, fish, penguins, and dolphins have each separately evolved flippers and fins as a solution to the problem of moving through water. What we had was something that has been less frequently demonstrated: *psychological convergence*. Dogs had independently evolved to be cognitively more similar to us than we were to our closest relatives.

The Truth About Domestication

Just about everyone thinks that domestication makes animals somehow weaker or less noble or just plain dumber, since our assumption is that humans created domestic animals for our own needs. People think of wild animals as noble and natural, and domesticated animals as artificial and engineered. It turns out that the truth is more nuanced, once we consider the origins of domestic animals.

Dogs, for instance, are not universally dumber than wolves. They have their own form of genius, one that – based on our first experiments with puppies and wolves – seemed to be a result of domestication. The most exciting part was that if we could test whether the social skills of dogs existed as a result of convergence with humans that occurred during domestication, we also might be able to figure out whether our social skills evolved due to a similar process.

There was just one problem with this idea – at the time, there seemed no way to test it. Without an experiment, we were slipping from science into the realm of storytelling.

Then, one evening in my second year of graduate school, our department was having dinner at a Chinese restaurant on Massachusetts Avenue. I overheard Richard talking about bonobos and how difficult it was to explain their evolution. He was talking about the psychological differences between bonobos and chimpanzees, where bonobos are more peaceful and less aggressive. There are also physical differences between the two species: Bonobos have smaller canine teeth than chimpanzees. Bonobos are more slender. They have smaller skull sizes.

'Oh,' I said, rudely interrupting the conversation. 'Bonobos sound like the domesticated silver foxes in Siberia.'

Richard turned politely and waited for me to explain myself.

'There's a breeding experiment in Russia where they bred foxes to be less aggressive. Over time the foxes changed, just as you were talking about with bonobos and chimps: small teeth, heads, and all that. It's in a chapter by Ray Coppinger from Hampshire College. He's one of the world's experts on dog behaviour and has studied dogs all over the world,' I babbled.

Richard stared.

'Can you get me that chapter by Monday morning?'

I gave Richard the chapter, and as a result, I ended up on a train to Siberia.

4

CLEVER AS A FOX

How an obscure Soviet scientist
revealed domestication's secret

Anyone familiar with twentieth-century Russian history would tell me Siberia was the last place I should be heading. Although the area was once known for its scientific intelligentsia, Stalin's three-decade rule left the field of biology in a shambles from which it has yet to recover.

When Stalin assumed power in 1924, it was clear his biggest challenge would be to stop the entire country from starving to death. Stalin's policies (including forcing peasants to work on common land to produce grain that was distributed by the government) created the worst man-made famine in history. Stalin knew that hungry peasants were angry peasants. No matter how many millions he sent to forced labour camps, there were millions more ready to stage an uprising unless he fed them.

What Stalin needed was a miracle of science. He needed crops to flourish at twice the speed with half the water. He needed fat heads of wheat to shoot towards the silvery winter sun; he needed potatoes to swell and fatten beneath the frozen soil.

Unfortunately, he turned his back on the one branch of science that could help him. Survival of the fittest, caricatured as the 'strong' thriving and the 'weak' perishing, smacked of the bourgeois oppression of the working class. So along with some of the worst crimes in modern history, Stalin turned his back on the new Darwinian science of genetics that would eventually revolutionize agriculture.

Let Them Eat Peas

And it was a truly new and exciting science. Although Darwin knew traits could be passed from one generation to the next, he was not sure how. The irony is that a copy of a scientific paper was reportedly found in Darwin's library after he died that could have solved everything.

Gregor Mendel was an obscure Austrian monk who became known as the father of genetics after his death. Like Darwin, Mendel was interested in inheritance and how characteristics were passed on from one generation to the next. However, when Mendel started his experiments in 1856, he knew nothing of Darwin or the theory of natural selection. *On the Origin of Species* would not be published for another three years.

Over seven years, Mendel cultivated around 29,000 pea plants. He chose different physical traits in the plants and began working on a mathematical model to predict which traits their offspring would inherit.

For example, two flowered pea plants may look exactly the same, but hidden in their genes is a dominant and a recessive bit of gene code, called an allele. When cross-pollinated, in each parent, two alleles split and only one is passed on to the next generation. So of four offspring, on average one would be grey with two dominant alleles, two would be grey with a dominant and a recessive allele (like their parents), but one would contain two recessive alleles, which would create a white flower.

And so you get the Punnett square many of us saw in school biology lessons:

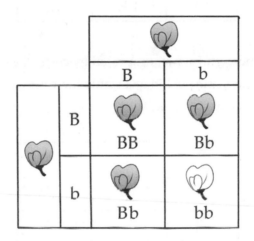

At first, no one knew what this had to do with natural selection. But say you add a selection pressure, so that in a certain field, the grey flower happens to be more attractive to grazing cows. Most of the grey flowers would get eaten, and the white flowers would have more chance of being pollinated. The next generation may look like this:

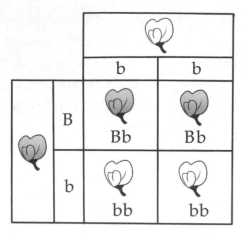

As even more grey flowers get eaten by cows, the probability of two white flowers being pollinated would increase, and further generations would look more like this:

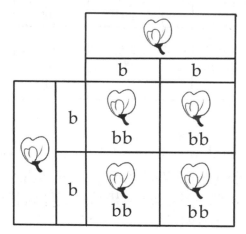

This is a simple example of how selection pressure can cause change to subsequent generations.

Some historians claim that a copy of Mendel's 1866 paper on his pea plant experiments was found in Darwin's library when he died.

At the time, pages were folded, and you had to slit them with a letter opener to read the pages. Darwin's copy was still uncut, which means he never read it.

Even if Darwin had read it, it is unclear whether he would have been able to understand the significance. Mendel used a heavily mathematical model, and it took the rest of the scientific world more than three decades to catch up. It was not until the 1930s, long after Mendel was dead, that the British statistician and geneticist Sir Ronald Fisher put Mendel and Darwin together in what became a grand synthesis.

Although Darwin could experiment with species that were already domesticated, such as pigeons (or peas, if he had thought of it), there was no data on the origin of domestic breeds. No one took any notes as the dog split from the wolf, or the pig from the wild boar. Domestication could completely be a product of *intentional* selection – the kind where you purposefully bred dogs with short hair to get more short-haired dogs. Or it could be in part a product of *unconscious* selection, where you bred for one trait, then got a whole lot of other traits by accident.

The Peasant Genius

With Darwin's dilemma still unsolved, Stalin had found himself a Soviet hero – Trofim Lysenko. Lysenko was the son of a Ukrainian peasant. He stumbled on an old technique called vernalization, where he could make seeds flower earlier by exposing them to low temperatures. Wheat, for example, needs a long period of cold before it flowers, presumably so the seed knows that winter has passed and it is safe to flower in the spring.

Lysenko announced that by freezing grain for two weeks, you

could make winter wheat sprout in the spring – this would solve the problem of grain perishing because there was not enough snow. Like a true mad scientist, he claimed to have invented the technique, which had actually been around for a hundred years. He also claimed that the grain he produced was more lush and fertile and would save millions from starvation.

Lysenko first described vernalization for Stalin in 1935, a couple of years after the massive famine of 1933. He claimed he could create new varieties of wheat in a fifth of the time that other scientists said was necessary and that he could increase the country's wheat production up to tenfold. Then he made the leap Stalin longed for, claiming the changes he made to seeds would be passed on to the next generation.

The experiments, of course, were a failure. To convince the government that he was living up to his promises, Lysenko began to falsify his results. He rejected several well-established genetic principles and eventually rejected the concept of the gene altogether.

Before Lysenko, the field of Soviet biology was incredibly robust and well respected. There is an old joke among American biologists that if you think that you have made a discovery, you can be sure a Russian has already discovered it and published it in some obscure journal in Cyrillic. Dmitry Ivanovsky was the first to discover viruses in 1892. Nikolai Koltsov came up with a double-stranded giant molecule a full twenty-five years before Watson and Crick first described the DNA molecule in 1953. Nadson and Filippov saw X-ray radiation cause mutations two years before the American geneticist and Nobel laureate Hermann Muller.

The Great Terror of 1937–38 cut short many careers, however. Stalin was convinced that the government, the military, and virtually every level of Soviet society were plagued by corruption and infested with spies. He initiated a massive purge that saw 1.3 million people arrested and half of those sentenced to death. The others were sent to forced the gulags. Not surprisingly, any scientist brave enough to point out that Lysenko's work was a sham was either imprisoned, tortured, or executed, several at the request of Lysenko himself, who had grown powerful, arrogant, and angry.

After World War II, the state of Soviet biology declined even further as Darwin was demonized. This was partly in reaction to Nazi ideology, which twisted Darwinism to justify racial extermination campaigns. The Nazis thought of Russians as subhuman, and during their invasion of the Soviet Union they burned villages en masse and subjected the local people to public executions, sexual slavery, and torture.

When World War II ended and the Soviet Union's relationship with America and Britain soured during the Cold War, everything coming out of the West was seen as indolent and wrong. Darwinism was seen as a justification for capitalists with superior strength or intelligence being rich while workers live in poverty. Genetics was seen as a tool of American imperialism to justify the racism that existed in American society.

As the West was slandered, everything Soviet was praised. Lysenko was made a hero, not just by Stalin and the Communist bureaucrats but also by the popular press, which hailed Lysenko as the embodiment of the barefoot scientist, a peasant genius. Photographs of Lysenko caressing ears of golden wheat against a blue sky saturated the Soviet media. A portrait of his heavy brow, clear eyes, and square jaw hung in all of the nation's scientific institutions. Monuments were erected in his honour.

Because everyone who disagreed with Lysenko was either fired, imprisoned, or executed, the next generation of Lysenkoites were almost completely uneducated but went on to occupy senior positions in the sciences, which set the field back decades.

Then, in 1948, came the ultimate triumph. Stalin supported Lysenko's call for genetics to be completely prohibited in the USSR. Genetic institutions were either closed or remodelled according to Lysenko's theories, and geneticists and their staff were dismissed. Genetic literature was banned in universities and removed from textbooks. Geneticists were officially declared enemies of the state.

And it was in this climate that one man would conduct perhaps the greatest behavioural genetics experiment of the twentieth century.

Darwin's Dark Knight

There is hardly any information on Dmitri Konstantinovich Belyaev. There are no biographies on his life, other than a few eulogies. After Belyaev's death, his wife published a book of memories from those who knew him, but this was distributed only among friends and colleagues, and it is impossible to obtain a copy. Most of the information we have on Belyaev comes from Lyudmila Trut, his protégée at the Institute of Cytology and Genetics, where she carries on his experiments to this day.

According to Trut, Belyaev was born in 1917, during World War I, in a small village called Protasovo, north-east of Moscow. True to the values of the time, Belyaev and his three siblings were taught to be good peasants; they harvested corn and took care of the livestock. But the family also valued learning, and the elder brother Nikolai went on to become a geneticist.

After the second grade, Belyaev was sent to Moscow to live with Nikolai and continue with his education. Belyaev came of age in what must have been an intellectually stimulating environment. As a prominent geneticist, Nikolai would have introduced his younger brother to his colleagues, and long afternoons must have been spent discussing exciting developments – and, in time, the persecution and decline that was to come.

In 1937, Nikolai was arrested by the secret police and shot without trial. Belyaev was twenty years old at the time. Belyaev did not return to his parents' farm. Instead, he stayed on in Moscow to pursue a dangerous path. The year after his brother was shot, Belyaev went to work at the Department of Fur-Bearing Animals at a state fur farm, essentially beginning his career in genetics.

His research was interrupted by World War II, and in 1941 he was drafted into military service. Beginning as a lowly machine gunner, Belyaev worked his way up to the rank of major. By the time he returned home, he had received several medals for bravery and service.

It is hard to do justice to how incredibly courageous Belyaev was to continue his work after the war. Although he was a war hero, this was no protection from persecution. Just before the war, Stalin had executed every army commander of the first and second rank. Returning soldiers like Belyaev were often suspected of having been corrupted by foreign influences.

Belyaev started his genetic experiments right as the gulag population was at its highest – more than 2.5 million people. He was treading a fine line. His 1946 dissertation, 'The Variation and Inheritance of Silver-Coloured Fur in Silver-Black Foxes', certainly sounds like the ideological heresy of Mendellian genetics. In 1948, when genetics was banned altogether, Belyaev was sacked from the Department of Fur Animal Breeding at the Central Research Laboratory of Fur Breeding in Moscow.

If Belyaev continued his work, it was not a possibility that he would end up in the gulags – it was a probability. In fact, he must have known that every day might be his last, that each night, he might hear the midnight knock on the door. Yet he continued his research.

In 1953, Stalin died. And with this, the chokehold Lysenko had on the field of genetics began to wane. But it died a long, slow death. The government after Stalin was openly supportive of Lysenko. The press was still pro-Lysenko. There were no degrees in genetics. If you wanted to publish an article in genetics, you had to do it in a chemistry, mathematics, or physics journal. It was not until more than a decade after Stalin died, in 1965, that Lysenko was forced out of the Institute of Genetics and the field of genetics began to recover.

By then, Belyaev's experiments were already extraordinary. Like Darwin, Belyaev was interested in domestication. Darwin did not come straight out with his hypothesis that humans share an ancestor with other apes. In *On the Origin of Species,* he started gently with a topic everyone was familiar with – selective breeding – to prove that selection could take place. Everyone was familiar with the idea that you could breed dogs to produce various characteristics. The same was true of pigeons, pigs, and other domestic animals.

Darwin referred to domestication as 'an experiment on a gigantic scale' in evolution. He used artificial selection, where breeders selected various traits to be passed on to the next generation, to illustrate the identical process of natural selection where, instead of a breeder, the struggle for survival drove evolution. But how did domestication happen in the first place?

And that was where Belyaev came in. He decided he would

domesticate his own species from the beginning. After a decade of dodging suspicion and wondering if he would share the same fate as his brother, in 1959 Belyaev moved to Novosibirsk, a remote outpost in Siberia, and became director of the Institute of Cytology and Genetics. He remained there, unmolested by the authorities, until his death in 1985.

The animal Belyaev chose to work with was the silver fox, because his great genetic experiment could be disguised as a commercial enterprise. In the north-western corner of Siberia, the pelts of these animals were the thickest and the softest, in snowy shades that were the ultimate luxury item. Silver foxes are a variety of the red fox (*Vulpes vulpes*), the most widely distributed of all the living foxes, ranging from the Arctic to deserts to the city. Although distantly related to dogs, foxes have never been domesticated. Silver foxes are those whose red-and-tawny fur has been replaced by black. The light guard hairs, the longest hairs on the coat, are responsible for the silver appearance.

Silver foxes have been bred on fur farms in Russia since the late nineteenth century. The main aim of the breeders was to increase the white guard hairs to emphasize the silver quality of the coat. However, one of the problems they faced was that patches of red and tawny yellow could pop up through the generations, reducing the value of the coats. Also, despite being farmed for generations, the foxes would still bite if you attempted to handle them.

Belyaev noted that a range of differences could occur between domestic species and their wild counterparts. Domestic species experienced changes in body size, leading to dwarf and giant breeds. Their coats became patchy in colour, and their hair could grow very long or become very short. Parts of their skin lost pigmentation.

Their tails curled. Also, while wild animals reproduced strictly sea-sonally, domestic animals could reproduce at various times of the year.

Usually, to select for traits, like a curly tail, you would breed two dogs who had curly tails. In order to create a domestic species, you could selectively breed animals for certain physical characteristics and hope domestication followed.

But Belyaev did something very different. Instead of breeding for many different *physical* characteristics, he bred for one single *behavioural* characteristic.

He started with thirty male foxes and one hundred vixens from a fur farm in Estonia. They had already been bred for fifty years, which meant they skipped the initial stressful period of being captured and forced to be near humans. Still, more than 90 percent of them were aggressive or fearful. Only 10 percent had a quiet, exploratory reaction to humans, without fear or aggression. Even these could not be handled, and bite precautions had to be taken. Belyaev called his initial fox population 'virtually wild animals'.

And then Belyaev added his selection pressure – how foxes reacted to humans. When fox pups were a month old, a human experimenter tried to touch and handle them. This was done every month until the pups were around seven months old. Each breeding season, Belyaev bred the foxes who were the least aggressive and most interested in humans. They formed a new experimental population.

Belyaev did one more crucial thing. He kept another group from the original population. This group was bred randomly as to how they responded towards humans – the foxes' behaviour towards humans was not a factor in their selection for breeding. This was called the control population. This way, he could measure any changes his selection criteria caused in the experimental population by comparing them with the control population.

After only twenty generations, the experimental foxes began to change in ways that might take thousands if not millions of years in the wild. By the time I arrived, the foxes had been bred for forty-five generations, and the experimental and control populations were radically different. Comparing the cognition of the two groups would test whether the key to the genius of dogs was domestication.

Belyaev's Time Machine

The train from Moscow to Novosibirsk takes two days. The Trans-Siberian Railway was built by the tsars in the nineteenth century and spans the country from the border of Finland to the Sea of Japan. Siberia is beautiful in the summer. We passed grasslands covered in flowers as the train brushed the tip of Kazakhstan, then headed east for Novosibirsk.

From Novosibirsk it is another hour and a half south to Akademgorodok – the scientific haven of Siberia. We drove six miles from the city and arrived at the fur farm. Very few outsiders have visited, and before I arrived in 2003, no foreigner had ever collected and published data on the foxes.

Belyaev was not allowed to share his extraordinary work with the outside world. He could not risk implicating or endangering his colleagues. Until the late 1970s, he mostly avoided foreign journals because academics who shared scientific information could be criminally prosecuted.

To stay out of trouble with the Soviet authorities, Belyaev claimed the main purpose of his research was to assist the nation's economy by breeding better foxes to improve the quality of their fur. But Belyaev must have known that if he was successful in domesticating the foxes, they would end up with splotchy colouring, star mutations, and pigmented skin – essentially rendering the pelts worthless.

However, Belyaev's cover was so effective that when the Soviet Union relaxed its policy on genetics, still no one knew what Belyaev had accomplished. I first heard of Belyaev when someone told me that a Russian farmer was breeding foxes for fur and had stumbled on some discoveries about domestication by accident. It was not until I read the chapter by Raymond Coppinger and discussed it with Richard that I realized the truth – the fox experiments were carefully masterminded by one of the greatest biologists since Darwin.

As I explored the farm, I walked by the control foxes who were not selected based on their behaviour towards humans. There was little variation in their coat colour, which was mostly black and silver, helping camouflage them during short summers and dark winters. As I passed, they melted away to the back of their rooms. Many of them began vocalizing, using the fox equivalent of a threat bark, which sounded like *chuff!* They were clearly not eager to interact with strange people.

Then I came across Belyaev's creations. Some had floppy ears, while others had curled tails. As I walked past, the foxes uncontrollably wagged their tails like puppies, pushing their noses against the bars, desperate to be petted. When I took these foxes out of their rooms, they leapt into my arms, nuzzled against my face, and licked my cheeks with their little pink tongues.

Behaviourally, there is a big difference between domesticated and tamed. You can tame a wild animal by raising them from birth or by feeding them. The wild animal you raise will never develop the normal fear of humans, while the wild one you feed will slowly lose some of their natural fear. However, the offspring of these wild animals will still be as wild as ever, since they still have their parents' genes. These tamed wild animals are not domesticated animals. Actual domestication involves genetic modification resulting in behavioural, morphological, and physiological changes that can be passed to the next generation.

It is not a matter of acting the 'right way' around wild animals or even just being with them all the time. Neither the experimental nor the control foxes had been raised with humans. In fact, both populations had very little contact with humans. Belyaev was very careful about this point. He did not want anyone to say the changes in the foxes occurred because they were tamed by humans during their lifetime and therefore their offspring were tamer. After being whelped, the only time foxes from either group saw a human was when they were fed.

Yet just like puppies, these foxes showed affection towards humans when they were barely weeks old, as soon as they opened their eyes. The team experimentally controlled for everything. They raised experimental pups with mothers from the control population and vice versa. They even invented a method for fox in vitro fertilization so they could implant the embryos of experimental foxes into mothers from the control line. None of these techniques reduced the experimental foxes' affinity for humans. The change observed in the behaviour of the experimental foxes was clearly driven by a genetic change caused by Belyaev's selection regime.

The genetic change plays out in the brains of experimental foxes. In experimental foxes, the levels of corticosteroids – hormones that regulate stress – were a quarter of those seen among the control foxes. The experimental foxes also had higher levels of serotonin, a neurotransmitter that makes humans feel happy and relaxed.

More surprising were the physiological and physical changes that happened 'accidentally' as a result of selection for the foxes' behaviour. The experimental foxes had more flexible breeding cycles. They reached sexual maturity a month earlier and had a longer breeding season than the control foxes. Their skulls were more feminized than the control foxes', resulting in shorter and wider muzzles, similar to differences observed between dogs and wolves. The experimental foxes were more likely to have floppy ears, curly tails, and splotchy coats. All of these features appeared as by-products of breeding foxes who were less aggressive and more social towards people. These are the same differences we see between domesticated animals and their wild ancestors.

Belyaev had done it. He had taken a population of wild animals and essentially domesticated them. And not just that, but he had figured out the mechanism by which it happened – not by intentionally breeding for each physical trait but by selecting only for behaviour, that is, by allowing animals to breed who were friendly towards people. All of the other changes associated with domestication occurred as a by-product.

Richard thought that cognition might be another accidental

by-product of domestication. If dogs had accidentally developed their extraordinary ability to read human gestures as a result of being domesticated, the experimental foxes should be more skilled at those human-gesture-reading games than the control foxes.

I thought the opposite would be true. I did not think that a change in cognition could happen by accident. I thought that it would probably require direct selection, so that cleverer foxes would have to breed to produce cleverer foxes. If I was correct, both populations of foxes would fail to understand human gestures.

The Vodka Solution

I sat naked in a Russian *banya*. The air in the sauna was so dry and hot, it scorched my windpipe all the way down to my lungs. Beads of moisture struggled to the surface of my skin and were instantly incinerated.

Eight Russian men, also naked, were leaning against the cedar walls, their eyes closed in ecstasy, as though slowly roasting yourself alive was the most relaxing thing in the world. Irene Plyusnina, the lovely host scientist who was helping us with our research, had sent me to the *banya* with her husband, Viktor, for an authentic Russian experience. Viktor had positioned me right beside the open fire, and every now and then the Russians chortled and muttered something about the 'Amerikan'.

To distract myself from the sensation that my eyeballs were being poached like eggs, I thought about the disasters that were my experiments.

I had been in Russia for two weeks with Natalie Ignacio, a Harvard undergraduate, and we were totally stuck. We had started by testing the experimental fox kits, who were between two and four months of age, because we could compare their performances with those of dog puppies. Like the dog puppies, the fox kits had almost no exposure to humans, so they could not have slowly learned to read human social gestures.

Irene brought out one of the kits. We let her sniff around the room for a bit, before Natalie centred her between the two cups. I showed her food then sham-baited the cups, touching both but only baiting one. Then I pointed towards the baited cup and waited for her to fail. Instead, she was awesome. She got almost every trial right. We tested a whole group of experimental fox kits from different litters. They flew through the tests, scoring high every time. We then ran the control test we used with Oreo to make sure foxes could not smell the food. Like dogs and wolves, the foxes all performed at chance levels. This showed they were not locating the food with their noses in this context. The experimental fox kits were not just performing as well as the puppies – they were even a little *better*.

Sometimes being wrong is more exciting than being right. Belyaev's domestication experiment might have also accidentally made the foxes more intelligent. But the only way to know for sure was to compare the performance of the experimental foxes with the control foxes. All of the foxes might just be more clever than dogs and wolves at understanding our gestures. If the control foxes could pass the test, then it was not domestication that made the foxes so clever.

And here was the problem: The control foxes were too shy to do the test. Since the Russians did not socialize the foxes, when we brought them to the testing room they were too nervous to be interested in the food. This made it difficult for them to make a meaningful choice. We had to come up with a fair test that would allow the control foxes to make choices confidently; otherwise we were toast.

I closed my eyes and leaned against the cedar walls. I yelped and sat up again. The wood was the temperature of a volcano, and my skin was taking on the colour and texture of a boiled crab. I had to get out. I stood up and tried to walk out nonchalantly.

'Brain. Wait for us.' It was Viktor, Irene's husband. Everyone in Russia called me 'Brain' instead of Brian. The large, naked men roused themselves like bears waking from a deep winter of hibernation. They shook off the heat clinging to their body hair and filed out of the sauna.

The sudden chill of the air sent tingles of fear rippling along the surface area of my skin, especially when Viktor gestured towards the plunge pool, which had the Arctic stillness of a glacier.

'Now,' he said menacingly, 'you jump.'

My head involuntarily started shaking from side to side. Viktor grabbed my arm.

'You Amerikans,' he said with a wry smile. 'In winter, we jump in the snow. This is unfrozen water. You must jump. You will love it.'

With my country's honour and my manhood at stake, I jumped. The shock was indescribable. Around me, jets of bubbles shot towards the surface as Russian men launched into the pool like concrete blocks. I kicked towards the surface and gasped.

The other men were already out and beating one another with birch branches. I decided to tread water to avoid being beaten by large Russian men with sticks. The cold felt like someone was peeling off my skin. When I could stand it no longer, I snuck past the men into the other room.

Like a magician, Viktor was already there, pouring a beer. Shivering, I accepted. I wrapped a towel around my waist and sat down on a chair, sipping slowly.

'No, Brain, no,' Viktor admonished. 'Beer without vodka is like money to the wind!'

The men gathered round the table, chasing beer with vodka shots. I was blind drunk after the first round, but Viktor was adamant and kept filling both my beer and my shot glass.

After what seemed like hours, the men got up. Thanking God my ordeal was over, I stood shakily and wobbled towards the changing room.

'Brain! Where are you going? We still have four more hours! Back to the *banya*!'

Four hours later, I was a mess. The dehydration of the sauna, the mixture of beer with spirits, the fact that I had not eaten since breakfast – all contributed to a very bad situation. Ironically, the only thing that stopped me from passing out was the freezing plunge pool. The water acted like a defibrillator, sending an electric shock through my heart and yanking me out of an alcohol-induced coma.

On the very last round, I sat in the sauna and leaned against the wall. I heard my skin sizzling, but I could not seem to feel anything. I closed my eyes and drifted up through the *banya* into the silvery summer sky.

A vision appeared before me. It was of a control fox playing with a feather that had fallen into his room. The control fox swatted at the feather with a black-gloved paw, his luxurious bushy tail swinging with lightness and grace. For once, even though I stood next to him, the amber eyes were not afraid.

I sat up. Where was I going to get a bunch of feathers?

Launching Sputnik

I now had a two-point plan. First, as soon as I got back, I told Natalie to start working with a group of control fox kits so that they would become highly socialized. The kits were between two and three months old, the youngest foxes on the farm. Every day for six weeks, she was to take them from their littermates and just hang out with them in a room for several hours. She would take baby steps to get them ready for the test. She would put a piece of food under a cup and see if they would approach and touch it. If they did, she would give them the food. Next, she would hide food under one of two cups and see if they would even make a choice, regardless of whether or not they were correct. If she could get them that far, we might have a chance to test them.

Second, in case Natalie could not socialize her foxes in time, I came up with a different version of the same test. What I had realized in the *bunya* was that even if the control foxes were shy of *people* at first, all of the foxes seemed to really love *toys*. Though the control foxes slunk away when I first approached, I found that if I sat quietly in front of their room, they approached me within a minute. If I wiggled a feather in front of them, they seemed to lose their fear completely. They would immediately approach the feather and begin chasing and playing with it as I held it in my hand. It was like Dumbo's magic feather. If I could find some way to incorporate the feather in the test, they might actually play the gesture game.

First, I needed a table. It had to be the same height as the fox rooms, and it had to have a top that slid towards the foxes so they could make a choice without getting too close to me.

I could have thrown the thing together in an afternoon, but Irene would not hear of it. This was a serious scientific experiment, and the equipment she provided would not be some shoddy American contraption, but a marvel of Russian engineering that Belyaev would be proud of.

The table project was sent to the workshop division, and they began to build the table with a seriousness that would have been more suited for a space mission. It took two weeks. I was almost out of my mind with impatience.

When the table arrived, my impatience was replaced with delight. Instead of the primitive plywood structure I had in mind, the table was sleek and modern, built from solid metal with a Plexiglas top that rolled forward without a sound. Attached to either side of the tabletop were the two matching toys I had created – Soviet red plastic boxes, each with a strip of metal tape measure stuck to the top. I could slide the tabletop so that the pieces of tape measure attached to each box would come within the foxes' reach.

The fox kits went nuts bending the tape measure with their paws and snouts so they could react as it sprang back into shape with a seductive clicking noise. The toys were like catnip for foxes. The table was beautiful. I christened it Sputnik, which drove the Russians wild with hilarity.

'Brain!' they would call out every morning. 'How is Sputnik?'

And now, here I was. We had less than half the trip left, and so far we had almost nothing. Natalie was playing with the fox kits as much as time would allow, but we had no idea if they could learn to play the cup game. This made my study with Sputnik all the more important. It might provide the only method to compare the two populations.

The control fox watched me as I positioned Sputnik in front of his room. He slunk in the back while I sat down behind it and set up the camera. He was very handsome. His silver coat shimmered, and his black ears twitched as he eyed me warily.

Everything changed when I pulled out the feather attached to the end of a stick. He immediately came forward and sat on the wooden plank that I had positioned in the centre of his room. He followed every movement of the feather (by the way, the experimental foxes were equally interested in the feather when I showed it to them). Curiosity overcame fear. I put the feather out of sight and touched one of the two toys while he watched.

The tape measure bent, making that irresistible clicking as it sprang back into shape. Next, the tabletop moved forward so the toys slid noiselessly towards him, both arriving within easy reach simultaneously. He immediately leapt at one of the toys and began flicking the tape measure and rolling on top of it playfully.

I was in business. Both the experimental and the control foxes loved to choose one of the tape measure strips to play with. I would slide the tabletop forward, touching one of the tape measure strips. I had a method to compare the two populations that required no socialization, training, or food rewards.

As a type of control, I also ran another version of the same test, but I hid my hand as I gestured towards one of the toys. I held up a board that hid me from the neck down to the tabletop and prevented the foxes from seeing my hand at any time. Then with a feather on a stick poking out from under the board, I touched one of the tape measures with the feather. Again, the fox leapt on one of the toys, swatting the tape measure back and forth.

The control foxes were no longer afraid to participate, even though I sat nearby. I could compare their preferences to those of the experimental foxes.

While both groups of foxes loved playing with the toys, they had completely opposite preferences. The motley crew of friendly experimental foxes preferred playing with the toy that I touched with my hand. The control foxes preferred to play with the toy that the feather on a stick touched.

The different preferences appeared on the very first trials and the foxes maintained these preferences after repeated trials, even though we did not reward them with food. This was a strong indication that Belyaev's selection criteria had altered how foxes treated a human gesture.

Meanwhile, after six weeks of socialization, Natalie also had success. She actually had a group of control fox kits that were comfortable around her and could make a correct choice when Natalie showed them she was hiding food in one of two cups. It was time to see whether they could use her gestures to find the food.

After another week of testing, and two days before we were to fly out of Moscow, we had our final results. The control foxes looked very similar to heavily socialized wolves and chimpanzees. They were not guessing randomly in every trial, but they were not very skilful at using our gestures. The experimental foxes were much better at interpreting our gestures to find the hidden food, even though they had had much less exposure to humans.

Both of our tests pointed to the same answer: Belyaev's experiment had changed the foxes' ability to read human gestures as a direct result of experimental domestication. Domestication, selecting the friendliest foxes for breeding, had caused cognitive evolution.

Richard was right, and I was wrong. The experimental foxes understood human gestures, even though the Russians had not bred them to be better at understanding human gestures. They were bred to be friendly towards humans, and like floppy ears or curly tails, they gained a better reading of human gestures by accident.

If we had tested the founding population of foxes when Belyaev started his experiments in 1959, they would not have read human gestures like the experimental foxes ended up being able to do. Those original foxes could respond to the behaviour of other foxes, but their fear of humans meant that when they saw a human, their first impulse was to run. The control foxes had the same impulses as these original foxes. But when we either socialized the control foxes by interacting with them for weeks or captured their attention with toys made of feathers or tape measure strips, their normal fear of humans subsided. When fear was replaced by curiosity about humans and toys, these foxes had some success using human gestures. This curiosity enhanced the control foxes' nascent skill at reading social behaviours. And this was a skill that was already present in the founding population – originating from the need to read the behaviour of other foxes.

Selective breeding in the experimental foxes had completely removed their fear of humans through genetic evolution. Fear had been replaced by a strong motivation to interact with us as though we were foxes. The change in their emotions allowed them to interact with us and solve problems that other foxes could not solve without intense socialization with humans.

Do-It-Yourself Domestication

The foxes totally rocked my world.

Before I went to Siberia, I subscribed to a more traditional concept of domestication, like that described by University of California at Los Angeles bio-geographer Jared Diamond: 'By a domesticate, I mean a species bred in captivity and thereby modified from its wild ancestors in ways making it more useful to humans who control its reproduction and (in the case of animals) its food supply'.

The fox experiment suggested the potential for natural selection to produce many of the changes originally attributed to humans intentionally breeding animals like wolves. If the least fearful and friendliest animals were at a natural advantage over more fearful and aggressive animals, populations with traits of domestication would evolve on their own without human control of their breeding.

Before Siberia, I was also sure that you had to breed the cleverest animals in a population to get a cleverer generation of animals. I thought that to create foxes who understand human gestures, you had to breed together foxes who were best at understanding human gestures. Instead, Belyaev had bred the friendliest foxes, who became cleverer by accident. Friendlier animals might naturally have an advantage over more fearful animals in settings where they needed to forage near humans and understand how to respond to human behaviour.

The foxes raised the real possibility that natural selection may have shaped wolves into the first proto-dogs in a similar way, without intentional human intervention or control. Ray Coppinger and others have speculated that, as humans began forming more permanent settlements over the last 15,000 years, a new food source appeared that led directly to the evolution of the dogs we know and love – rubbish.

Wolves who usually avoided humans (like the control foxes) were attracted to piles of bones, human faeces, rotting meat, and starchy vegetable scraps. Wolves who were too afraid to approach humans would not be able to take advantage of this new ecological niche. Wolves who were willing to approach but were too aggressive would have been killed by humans. Only the wolves who were least fearful and non-aggressive towards humans would be able to take advantage of this new source of food. Like the foxes, they, too, accidentally became more skilled at responding to the behaviour of humans.

The first few generations of wolves may have silently approached under the cover of darkness. With a more stable food supply, they would have more surviving offspring. These offspring would then inherit their parents' more relaxed genetic predisposition towards humans. This cycle would then repeat itself over generations as these less reactive wolves introduced their calmer offspring to this new method of foraging around human settlements.

It would not have taken many generations for these friendlier wolves to undergo physical changes (changes in coat colour were already occurring in the eighth generation of foxes). Soon, the wolves stopped looking like wolves. Many would have splotchy coats, and some would have even had floppy ears or curly tails. Initially, humans might not have been too welcoming towards these bold wolves, but to the wolves, the benefits of feeding on rubbish piles would have outweighed the penalties of being chased off, harassed, and occasionally killed.

With morphological changes appearing after only a few generations, people would quickly be able to recognize this new type of wolf or proto-dog. Like many modern societies, these first 'village dogs' could be ignored, occasionally eaten, or even taken as pets when young. Humans did not set out to domesticate wolves. Wolves domesticated themselves. The first dog breed was not created by human selection or breeding but by natural selection.

At least that was the idea the foxes gave us; now I wanted to find a way to test it directly with dogs.

New Guinea Singing Dogs

The foxes showed that friendly individuals tended to have offspring who were not only friendly but also able to read human gestures. If selection against aggression causes domestication, and if dogs domesticated themselves, then the earliest dogs would have been skilful at learning from human gestures, before humans started intentionally breeding them.

What we needed was a modern version of these early dogs, who had not been intentionally bred by humans.

Feral dogs are domesticated dogs who have gone wild. They may ride the Moscow underground, roam the streets at night, or live in a national forest and scavenge on rubbish dumps. What all of these dogs have in common is that they do not have a human family. Similar to the first proto-dogs, they still depend on scavenging near human settlements. But unlike the proto-dogs, feral dogs recently descended from pet dogs, so they carry the genes of dogs who have been intentionally bred by humans.

The two exceptions are the dingoes of Australia and New Guinea Singing Dogs. Both can be socialized like a domesticated dog, look very dog-like, and are genetically closely related to Asian breeds. However, these dogs probably went feral as far back as 5,000 years ago and live in wolf-free environments; researchers suspect that neither was intentionally bred by humans.

Dingoes and New Guinea Singing Dogs are probably the closest modern representative of proto-dogs. If proto-dogs did domesticate themselves, then dingoes and New Guinea Singing Dogs should also be skilled at reading human gestures.

Out of convenience, I decided to test New Guinea Singing Dogs. I did not have to travel as far as New Guinea. Along the Rogue River, in Eugene, Oregon, Janice Koler-Matznick runs the New Guinea Singing Dog Conservation Society, where she cares for a population of New Guinea Singing Dogs. Endemic to the alpine mountains of New Guinea, these dogs can live at altitudes up to 15,000 feet (four kilometres), which is almost a thousand feet higher than the highest point of the Rocky Mountains. New Guinea Singing Dogs are the only canid species besides the Ethiopian wolf that can live at such high altitudes. They are amazingly cat-like and can climb trees in search of prey they occasionally steal from harpy eagles.

They are also the most risqué of the canids — they regularly masturbate and tend to bite one another's genitals, both playfully and aggressively. During copulation, the females emit a high-pitched yelp for three minutes at a time, which has an arousing affect not just on other New Guinea Singing Dogs but on all of the domestic dogs within hearing distance. They are also one of the rarest dogs in the world.

When my colleague, Victoria Wobber, and I arrived, we were greeted by the eerie sound of the dogs singing. Each dog can sing a different note and hold it for up to five seconds before starting again, creating a chorus that has been described as half wolf howl, half whale song.

When the dogs saw us approach, they gave a weird head toss – a unique behaviour of New Guinea Singing Dogs, where they flick their heads from side to side and occasionally turn a full circle. They were absolutely mesmerizing but at first a bit wary. After a day or two, they warmed up to us, and we were able to teach them the basics of being tested. Then, as always, we hid food in one of two cups and tried to communicate where to find it.

We tested their ability to use three different gestures: pointing and gazing, placing a block, and placing the block while the dogs had their eyes covered. The New Guinea Singing Dogs aced the test every time. They were skilled at reading our gestures, even though no humans had bred them for this purpose. Recently, dingoes have shown similar skills.

These half-wild proto-dogs seem to demonstrate that the ability to read human gestures arose early in domestication and did not require human selection. Our results also supported the idea that proto-dogs domesticated themselves. This was the final piece of the puzzle. Humans did not create dogs; we only fine-tuned them later down the road.

A Blueprint for Domestication

Through dogs, wolves, chimpanzees, and foxes, we now have a clear picture of the initial stages of domestication and its effect on social skills.

Our remarkable relationship with dogs began when populations of wolves took advantage of a food source near humans. As humans encountered these self-domesticated animals, they would have realized that these proto-dogs responded to their gestures and voices. These same dogs, as they lived on the outskirts of a human settlement, would have begun barking when strangers approached, providing an early warning system. In times of famine, they may also have been a crucial source of food. Like ravens following wolves to a kill, these proto-dogs would have begun shadowing human hunters to scavenge left-over scraps discarded when butchering occurred at the kill site.

Just as modern Hadza hunter-gatherers in Tanzania are led by honeyguide birds to beehives full of honey, humans could have easily begun paying attention to when these proto-dogs began chasing or barking at prey. With their projectile weapons, humans could then finish the job when the dogs cornered their prey. All the while, selection continued to favour the dogs who were friendliest towards humans. In experimental tests, dogs look similar to the experimental foxes in preferring the company of humans to that of other dogs – while even wolves raised by humans prefer other wolves over humans. This attraction to humans helped dogs move from the outskirts of human settlements to the rug by the hearth inside our homes.

All this leads to a much bigger question. What if something similar could happen to other species? If natural selection can lead to self-domestication, perhaps other wild species self-domesticated – including humans. Many people have suggested that human cognition became so sophisticated because the most intelligent people survived to produce the next generation. But maybe it was the friendlier people who had the survival advantage, and, like the dogs and foxes, these people became more intelligent by accident. Could the self-domestication of dogs teach us something about human nature?

This was the last step in a scientific journey Oreo started me on years before. To find out the answer, I would have to cross the globe again, travel deep into the Congo Basin, and rediscover a long-lost relative.

5

SURVIVAL OF THE
FRIENDLIEST

How a little congeniality can get you ahead

Hey,' whispered someone behind me. 'It's that dog guy, what's his name?'

'Brian Hare?' whispered his companion. 'I think it's Brian Hare.'

We had just taken off from a small town called Mbandaka in the Democratic Republic of the Congo. The forty-year-old single-engine plane was a little worse for wear, and I heard the leather seat squeak as one of the voices tapped me on the shoulder.

'Excuse me,' said a man in his late fifties, sitting with his wife and another couple. 'Are you that dog guy?'

Whenever someone calls me 'that dog guy', I always think they must have me confused with someone famous. I used to explain that I'm an evolutionary anthropologist who studies many different species in order to understand human cognitive evolution – but that seems to make people look bored.

'Sure,' I say. 'Pleased to meet you.'

The couples were members of the American Kennel Club, and

they were flying to the Congo Basin to search for basenjis. Basenjis are dogs found in West and Central Africa that probably originated somewhere in the Niger-Congo region. Genetically, basenjis are among nine living breeds that are more wolf-like than all other breeds. Basenjis are rare in the US, and these couples were looking for puppies to take back to America to start a fresh bloodline. They had heard that some forest communities near where we were heading were breeding basenjis.

We talked for a while about basenjis and their wolf-like genetics. I was just about to tell them what I was doing in Congo when out of the corner of my eye, I saw that we were flying directly into a storm front. It looked nasty, with billowing grey clouds that swooped towards us. I tightened my seat belt.

The plane tilted madly from side to side; then we went into free fall. We pulled up only to fall again. After a few flashes of lightning and deafening thunder, it became very dark. With a final shudder, we burst into the sunlight. Trying to appear unruffled, I unclenched my grip on the seat and looked out the window.

Beneath us was a forest so ancient, so vast, that we could have been in the Land Before Time. The leaves created a green haze that met the blue curve of the horizon. There was no trace of human activity. Not a cleared acre or even a smoke plume. As we flew, the only thing that broke the jungle was the mighty river, snaking its course towards the ocean. From up high, the river reflected the blue of the sky, but as we dipped closer, its depths were as black as coal from the tannins that leached from the trees.

All I ever hear on the news is that the forests are vanishing; that there is no wilderness left on Earth. I never thought it could still exist, this ocean of trees. It was the most beautiful sight I had ever seen.

And deep within its canopy were the long-lost relatives I had come so far to find.

The Heart of Africa

Ten million years ago, movement in the Earth's crust pulled apart two tectonic plates in Central Africa. This created a shallow depression in the heart of the African continent that became the Congo Basin. Nearby, Lake Tanganyika would periodically flood, creating a large inland lake. The evidence of this lake is the floodplain we see today.

Around eight million years ago, mountain ridges raised the East African Rift Valley, drying out the savannahs to the east. It was around this time that the first primitive hominids emerged from the forests, beginning to walk upright to explore their new habitat.

To the west of the mountain ridge, the forest remained vast and unbroken – the perfect habitat for tree-dwelling apes. Over the years, the forest expanded and shrank, but the Marungu Mountains, on the shores of Lake Tanganyika in south-eastern Congo, would have been where these early ape ancestors sheltered while the forest contracted during periods of severe drought.

Over time, the waters of the Atlantic Ocean were channelled inland, where eventually they met the overspill from Lake Tanganyika. This formed the mighty River Congo, which would become the deepest river in the world, at times plunging to 750 feet (225 metres). The river arched like a rainbow towards the north of the country and, as it cut through the landscape, separated the early ape ancestors into distinct populations.

To the north of the river, the plates forming the East African Rift

Valley continued to separate, creating the right conditions for an enormous forest and a large transition zone to open woodland and then to savannah. It was this habitat that saw the evolution of these early apes into gorillas and one of our two closest living relatives, chimpanzees.

The south of the River Congo is a different story. A dramatic change in elevation from the deep basin to the highlands created a more specialized environment. There was almost no transition zone of open woodlands, which means the dense tropical forest became an isolated ecosystem. Many of the species that evolved in this environment would be extremely diverse, and some would exist nowhere else in the world.

One of these species was our other closest living relative, one who has remained a mystery for the better part of two centuries.

The plane landed on the hot, dusty airstrip of Basankusu in the Équateur province of Congo. Basankusu is a city of approximately 100,000 inhabitants with no electricity, no running water, and no resident doctor. I disembarked and resisted my urge to kiss the ground. I was travelling with the internationally renowned conservationist Claudine André, a handsome woman in her early sixties. Claudine was born in Belgium but has lived most of her life in Congo. When I asked her how she survived the flight, she coolly replied, 'Ah yes, it is the rainy season after all.' She had slept the entire way. Having lived through several wars and dictatorships, she was adept at surviving all kinds of turbulence.

We arrived at the Lopori River and got into a canoe that was a hollowed-out tree. Our driver pulled the outboard motor and we began to chug along at a good pace, pushing through water that was as still as glass.

Soon we left behind all trace of human habitation. Occasionally we passed a fisherman balanced barefoot on the edges of his canoe, throwing a net into the water. Mostly we were alone. On a bright day like this one, the river was a perfect mirror of the clouds in the sky, thick green trees, and birds flying overhead. The foliage was so dense, it spilled out onto the river, and branches bent with the weight of their leaves to soak in the water.

Miraculously, a beach appeared, the kind you would expect to see in the Bahamas. The sand was white and fine, strewn with dried palm fronds. We pulled up and dragged the canoe onto this unlikely shore.

A black form shimmied down a tree trunk and led her family onto the beach. They were bonobos – our closest but almost forgotten relatives.

Etumbe, the group leader, came first. She had a new baby on her back and was holding her son by the hand. When she saw Claudine, she shrieked happily and sat down close enough that Claudine could see two wide eyes peeping out from her mother's black fur. Beni somersaulted out of the forest laughing, while Lomela chased him, trying to grab his feet. Several more followed, until there were nine altogether. Claudine has known them all their lives. They were the Lola ya Bonobos. These orphans of the bush-meat trade had been released almost a year before I visited. It was the first time anyone had released bonobos back into the wild, and I was one of the scientists who had advised on the process. Claudine had brought me to see them in their new home.

We spent the day with the bonobos on the beach, cooing over Etumbe's new baby, watching Beni and Lomela play with the others. It was a delight to see them living happy, healthy, and free.

We stayed for a week and watched the bonobos each day. Everyone had worked hard to get them back into the forest, so it was an incredibly fulfilling experience. On our final day, we did not want to leave. When it was time, and we started heading from the forest to our canoe, Etumbe walked towards us in a determined manner. Knowing she was stronger than all of us, I was a little alarmed at her pace. As each of us passed her to board the canoe, she looked deep into our eyes, took each of our hands, held it in her own, and shook it in a gentle good-bye. As a scientist, I have no idea what she was thinking, but as a person it was the most heartfelt thank-you I have ever received.

Peace-Loving Jungle Hippies

Jane Goodall startled the scientific community when she discovered wild chimpanzees using tools and hunting monkeys. She was the first to begin uncovering these primates' rich social networks, which are dominated by friendships and strong family bonds. But the discovery that changed our understanding of animal societies came when Goodall witnessed the males of the Kasakela chimpanzee community systematically kill off members of the neighbouring Kahama chimpanzee community. As they killed more members of the Kahama community, the Kasakela chimpanzees slowly took over parts of Kahama territory. She had discovered that, like humans, chimpanzees have a dark side.

After decades of work in dozens of field sites across Africa, we know that Goodall's observations are typical of the species. Chimpanzees are universally hostile towards strangers. Gangs of male chimpanzees co-operate in patrolling their territory borders, op-portunistically kill their neighbours, and subsequently take over their territories. These gangs typically target males and infants when they attack. Females are normally spared but will later have little choice but to join the attackers' group if their own males lose too much territory. Chimpanzees are so skilled at working together to hunt down other chimpanzees that lethal intergroup aggression, or murder, is among the leading causes of mortality in wild chimpanzees. Rates of lethal aggression in chimpanzees are similar to the levels in some pre-agricultural human societies.

Chimpanzee aggression is not just directed towards strangers. As soon as males become adolescents, they beat up every female in the group until she submits, usually starting with their mother. Males then begin working their way through the male hierarchy. All chimpanzee groups have an alpha male, and it is the aspiration of every male to reach that position. The main prize for alpha males is sexual control over the females. Unfortunately, males often force females to mate with them and prevent them from mating with other males by severely biting and beating the females.

Genetically, bonobos are almost identical to chimpanzees, but they do not have this dark side. No one has ever seen male bonobos patrolling borders, raiding neighbouring groups, taking over an-other group's territory, or killing other bonobos. When neighbour-ing bonobo groups meet, it can be a big social event. When they meet, two groups will often stay near each other playing and having all types of sexual interactions. They may even travel together for days as one giant bonobo group.

The highest-ranking bonobo in a group is always a female, and females are close friends. Male bonobos do not beat up their mothers; indeed, they remain close to their mothers throughout their lives. Male bonobos do not use physical aggression to control females. Instead, a male bonobo's mother will simply introduce him to her friends. The males whose mothers have the strongest social network will have the most success in the bonobo mating game.

Though bonobos and chimpanzees are descended from the same ancestor, something happened that made bonobos less aggressive. The dogs had taught us about the effect of domestication on psychology. The foxes had shown us that domestication is a result of selection against aggression. Bonobos raised the possibility that nature can domesticate without any aid from humans.

The Dog of the Apes

Almost a century ago, a famous anthropologist and conservationist named Hal Coolidge was riffling through a box of bones in the Tervueren museum in Belgium. He picked up a small skull that looked like it belonged to a juvenile chimpanzee. Coolidge might have left the skull and moved on, but something caught his eye. In juvenile apes, the plates of the cranium, or braincase, are not fully fused. This leaves fissures in the craniums of animals who die as juveniles. But the fissures in the cranium of the juvenile-size skull in Coolidge's hands were fused, which meant it belonged to an adult. This had never been observed before, so Coolidge inferred that he was holding the skull of a completely new species. He published his findings in 1933, making bonobos the last of the great apes to be discovered.

One of the most distinctive features of bonobos is the cranium, which in males can be as much as 15 percent smaller than the cranium of a male chimpanzee. Bonobo skulls are 'frozen' in a juvenile state. This might sound strange, but there is one other group of animals that displays the same phenomenon: domesticated animals have a smaller cranium than their wild-type ancestors. Dogs have craniums that are 15 percent smaller than a wolf's of the same weight. Guinea pigs have skulls that are 13 percent smaller than cavies', and even domesticated fowl show a similar pattern. Smaller craniums made for holding smaller brains are a telltale sign of domestication.

Smaller craniums are not the only similarity between bonobos and dogs. When it comes to aggression, the differences between chimpanzees and bonobos mirror those between wolves and dogs. Like chimpanzees, wolves are extremely territorial. For instance, between 39 and 65 percent of wolves in Denali, Alaska, were killed by other wolves, usually along territorial boundaries of two different packs.

Both male chimpanzees and wolves become extraordinarily aggressive when competing for sexually receptive females. Male and female chimpanzees have been known to kill infants of other chimpanzees, while female wolves attack other females and occasionally kill their puppies. Both wolves and chimpanzees hunt, with wolves depending on hunting for food, while chimpanzee hunting can reach levels so high that entire species of monkeys are nearly extinguished within a group's territory.

Dogs and bonobos show much less severe aggression than their counterparts. Feral dogs have very rarely been observed to inflict mortal wounds on another dog. Instead of using physical aggression, dog packs usually work things out by barking at each other until one pack decides to leave. Feral dogs also have not been observed to kill the puppies of other pack members. As we have all seen, dogs are so tolerant they even let strangers sniff their bottoms.

Wolves do not tolerate this even from their own fellow pack members.

Adult dogs are more playful than adult wolves, and play like juvenile wolves throughout their lives. Bonobo adults play with each other the way juvenile chimpanzees play with adults, initiating more play and using more play faces.

Also, feral dogs show more sexual behaviour, including mounting and mating outside of reproductive seasons. As for bonobos, promiscuous, non-conceptive sex is what they have become known for. Their sex lives make chimpanzees and even humans look dull. Feral dogs are poor hunters while bonobos rarely have been observed hunting (in fact, bonobos may be as likely to play with monkeys as they are to try to hunt them).

Physically, bonobos are more slender than chimpanzees and have smaller canine teeth. Although bonobos do not have splotchy coats like dogs, they can have pink lips as a result of loss of pigmentation around their mouth, and a white tuft tail (chimpanzees lose this when they become juveniles).

The differences between chimpanzees and bonobos are strikingly similar to the differences between wolves and dogs. But bonobos have not had much to do with humans, at least in comparison with dogs. Why did they become so dog-like?

Ape Self-Domestication

A lot of different evolutionary explanations would be needed to make sense of these bonobo traits – friendliness, sexual and play behaviour, small cranium and canine teeth – if each trait evolved independently. Some characteristics do seem explainable on their own, while others do not. For example, heightened sexual behaviour relieves social tension and reduces aggression, but it is unclear how a smaller cranium or canine teeth would improve bonobo survival. The most likely explanation for all of these dog-like traits relies on understanding why bonobos became less aggressive in the first place.

When Richard Wrangham learned of our fox results, he realized there might be an explanation for how bonobos became so dog-like.

Belyaev's foxes showed that selection against aggression can cause a domestication syndrome. Humans bred the least aggressive foxes for several generations and created foxes who were friendly towards humans. These same foxes saw many changes occur by accident. No one selected for feminized cranium, smaller teeth, floppy ears, or skills at using human gestures, but these traits, all signatures of domesticated animals, appeared at high frequency in the friendly foxes.

We usually think of natural selection as the survival of the fittest – where the fit are those who are strong and aggressive, while the weak perish. In biology, however, *fitness* refers to those who are most successful at reproducing, not necessarily the most aggressive. A perfect example is human domestication of a species, where selection against aggression can increase reproductive success.

Natural selection can do the same thing. Our hypothesis was that when wolves spent more time near humans, without being intentionally bred by humans, the least aggressive wolves were the most likely to survive. Natural selection favoured the less aggressive wolves, and as with the foxes, accidental changes also occurred, turning them into the first dogs. Other wild species may have experienced selection against aggression. Over several generations, less aggressive individuals in these species would change in appearance, development, physiology, and cognition, similar to Belyaev's foxes. Essentially, a wild species could domesticate themselves.

Perhaps this is what happened with bonobos. The main push towards self-domestication may have been caused by bonobos having more reliable access to food than chimpanzees. There is some evidence that fruiting trees bear more fruit more often in bonobo forests when these have been compared with chimpanzee forests.

But more importantly, bonobos do not have to compete for food with gorillas, since gorillas do not live south of the River Congo. When fruiting trees are not available, chimpanzees must vie with gorillas for the low-quality ground herbs that gorillas regularly eat. This makes it very difficult for chimpanzee females to stay together on a daily basis. It is difficult to be social when there is little food to share. With no gorillas to compete against for this lower-quality food, and more fruit to start with, bonobo females can afford to be very gregarious. Bonobo females form strong bonds not observed in the less social female chimpanzee.

These tight bonds between female bonobos are the secret to their success. Although one female is less powerful than any male, if one female is being harassed, all of her female friends will come to her defence. In this way, females work together to protect one another from male aggression. Male bonobos can no longer force females into mating with them. Female bonobos have far more freedom in choosing with whom to mate as a result. Instead of bullies, females prefer to mate with the more gentle and peaceful males. The main reason to be an alpha male is to monopolize matings with the females in the group. If beating females and being aggressive no longer leads to reproductive success, there is no longer an evolutionary advantage to intense male aggression. Instead, friendly males will be the most successful.

Given that animals are always competing, it seems paradoxical that evolution would ever favour friendly animals. Richard's explanation for bonobos solves this paradox. By being less aggressive, male bonobos were more successful at reproducing. Also, since selection against aggression altered the morphology of the foxes, this one hypothesis could also explain the strange physical traits of bonobos, such as their smaller teeth and craniums.

All this goes to show that often, it's survival of the friendliest.

Meanwhile, Back in Congo

So now I had to figure out how to get bonobos and chimpanzees to play the same experimental games. Not only did I not have any Russians around to build me tables, but finding a captive population of bonobos to play cognitive games is like finding a baby dodo.

After flying through a thunderstorm to visit the released bonobos, we then had to fly back over the Congo Basin to where it all began. Lola ya Bonobo is thirty hectares of tropical forest near the capital city of Kinshasa. The tree canopy rustles as long arms swing gracefully from branch to branch. High-pitched cries fill the air. Bonobos forage for water lilies in the lake, dangle their legs from high branches, and do headstands in soft piles of moss. For the sixty orphan bonobos who live here, it is paradise, and that is what *Lola ya Bonobo* means in the local language – Paradise of the Bonobos. The woman who began it all is Claudine André, my red-headed companion who slept through the flight to Basankusu.

In 1997, Claudine saved an orphan bonobo named Mikeno. Bonobos are highly endangered because they are illegally hunted for the bush-meat trade. Mikeno's mother had been killed by hunters who had tried to sell Mikeno on the more lucrative, multibillion-dollar international exotic pet market.

Since Mikeno, Claudine has dedicated her life to protecting the last bonobos. She has worked tirelessly with the Congolese government to stop the traffic of orphans for the pet trade. Claudine created Lola ya Bonobo sanctuary for these orphans, the only bonobo sanctuary in the world, with the largest captive population of bonobos.

Here Richard and I, along with Victoria Wobber from Harvard University, did the experiments to test our self-domestication hypothesis.

We confirmed that bonobos are far more tolerant of sharing than chimpanzees. Two chimpanzees or two bonobos entered a room with food. The chimpanzees usually avoided each other, and one would try to take all of the food. Bonobos were different. When two bonobos entered the room, regardless of their ages, they always shared the food. They even played together while they were eating. The unusual sharing behaviour we observed in the bonobos also seemed to be juvenilized behaviour. Juvenile chimpanzees are much more tolerant of sharing than adult chimpanzees. Meanwhile, bonobos of all ages, including adults, were sharing at levels similar to juvenile chimpanzees. It was as if bonobo adults were 'frozen' as juveniles. We suspected that, like domestic animals, in some ways bonobos never grow up.

To understand the physiology behind this behaviour, we needed some ape drool. Saliva contains hormones that animals, including humans, release when they are stressed. But no one had previously collected saliva from a bonobo or chimpanzee. Victoria came up with an ingenious method. She crushed up some SweeTart candies and dipped a cotton pad in the powder. At the sight and smell of the confectionery treat, we had dozens of ready volunteers holding their mouths open wide and drooling. All we had to do was swab their mouths with the cotton and soak up their saliva while they sucked down the candy powder.

In the experiment, we released either two male chimpanzees or two male bonobos into a room with food to see if they could share. Before and after they entered the room, Wobber took samples of their saliva. The results revealed any change in the level of cortisol and testosterone during the sharing test. Cortisol increases when an animal feels stress. Testosterone increases in competitive situations.

Comparing the saliva samples from before and after the test confirmed our prediction. Male chimpanzees had an increase in testosterone when forced to share with another chimpanzee. This suggests that they perceived the test not as a chance to share, but as a competition that their bodies must be primed for in order to win. Bonobos, on the other hand, had no increase in testosterone but a marked jump in their cortisol levels. This suggests bonobos did not perceive the test as a competition to be won, but instead as a stressful social situation. Like a nervous couple on a blind date chattering to overcome their awkwardness, bonobos play and even embrace to overcome the same type of stress. The physiology of bonobos is not set up for competition over food that might lead to aggression. Instead, bonobos react in a way that prevents conflict and promotes sharing, which supported our prediction.

Another prediction of our bonobo self-domestication hypothesis was that just as the experimental foxes were non-aggressive and attracted to humans, bonobos should be non-aggressive and attracted to strangers. To test this idea, we gave bonobos a choice between sharing with a stranger and sharing with a group mate.

A bonobo entered a room with a pile of food. This room had another room on each side, one with a stranger, the other with a group mate. The bonobo could choose whom to let in.

Amazingly, the bonobos did not eat all of the food. They instead chose to open the door for strangers and share their food. They even preferred to let in the stranger over their group mate. Bonobos are definitely attracted to strangers.

All of our work supported the idea that the behaviour, development, and physiology of bonobos had made them a friendlier species than chimpanzees. The final challenge for the self-domestication hypothesis was to assess whether bonobos, like dogs, had also become 'smarter' as a result of their friendliness.

We already had the perfect test from our work with chimpanzees. In front of two chimpanzees is a long red plank, 1.8 metres long. At each end, there is a dish full of food. Next to each dish is a metal loop. A long rope threads through the loops so that in order to pull the plank, two chimpanzees have to pull both ends of the rope at the same time. If only one chimpanzee pulls, the rope comes unthreaded from the plank like a shoelace. To succeed, two chimps have to co-operate and pull both ends of the rope at the same time to drag the plank towards them and get the food.

Most of the time, the chimpanzees were extraordinary. They spontaneously solved the problem on the first attempt. They knew

when they needed help and who was better than someone else at co-operating. Chimpanzees collaborate at a level of complexity that is seen in human children.

But even though the chimpanzees were impressive, they had several limitations. First, we could test only a few pairs because, as we saw from the previous sharing test, most chimpanzees were incapable of sharing food. Most dominants get upset if a subordinate comes near the food – even if it is out of reach. This left the subordinate too frightened to approach the rope or the plank, making co-operation impossible.

Second, chimpanzees cannot share food unless it is divided into separate piles. All we had to do to make chimpanzee co-operation fall apart was put the food in the middle of the plank. With the food in the middle, one chimpanzee ended up hogging the food and the other would refuse to play the game. Even though the same two chimpanzees had co-operated successfully dozens of times before, having one pile of food made it impossible.

At Lola ya Bonobo, we gave Kikwit and Noiki the same test. Unlike the chimpanzees, who were experts, the bonobos had never seen the test before. As soon as Kikwit and Noiki saw the food, they raced forward and each pulled a rope, dragging the food within their reach. Next they did something we had never seen in chimpanzees: both bonobos reached into a pile of food and each took out a piece at the same time. They took turns taking food until it was all gone.

Even when we moved the food from two piles into one pile, the bonobos still co-operated, each taking exactly half of the food. All the bonobos we tested were the same. No fear, no intolerance; just successful co-operation, sharing, and play. Even when we put the bonobos with new partners, they could still co-operate. Bonobos

111

had beaten the chimpanzees at the same co-operative problem, even though the chimpanzees had had more practice. The bonobos' increased friendliness and tolerance made them more flexible co-operators than chimpanzees.

Just as with dogs, self-domestication has made bonobos smarter.

If natural selection can shape dogs and bonobos to be more friendly and more socially skilled, what about other species? The answer may be living in your neighbourhood. Dozens of wild species are repopulating areas where they were displaced when large cities and suburbs were rapidly built in the past century. This requires these animals to be less afraid of humans than their counterparts living in remaining wild areas. Over the last thirty years, the Florida Key deer has increasingly encroached on urban areas. Populations living nearest urban areas have become less fearful, bigger, and more social; they have offspring more often, thus showing higher fitness than deer living farther away from urban areas.

In addition to herbivores, carnivores are also candidates for self-domestication. Most notably, foxes, coyotes, and bobcats have begun taking up residence in urban areas. We may be at the beginning of the next great domestication event in the decades or centuries to come, where a number of carnivores will join the ranks of dogs by becoming self-domesticated as they adapt to life in the big city.

Self-Domesticated Humans?

As an anthropologist interested in human evolution, Richard could not help but tempt us to start thinking about how self-domestication may have occurred in our own species. It quickly became clear that self-domestication may be particularly helpful in explaining the evolution of our species's social cognition, which makes us so different from other apes.

Research on infants is what got us thinking about social skills in animals in the first place. Mike and other infant researchers recognized that the foundation of human cognition is the social skills that develop in an infant's first year. These skills allow infants to communicate and learn from adults in ways that other species cannot. At the same time, infants are unusually motivated to co-operate. Even as early as fourteen months of age, infants spontaneously help others and enjoy co-operative games.

This social revolution occurs very early in infants compared with apes. The skills that develop in infants allow them to take advantage of their tolerant environment. We adults share everything with our infants, from food and shelter to plush animal toys and books, and this nurtures their desire to be co-operative. This fact of life has been closely observed across cultures. Then, as we grow up, we mature into the co-operative fabric of a larger social group. Building the Great Wall of China, decoding the human genome, and forming a democratic government all require a level of co-operation that is unique to our species. But the question remains: How did humans evolve to be more co-operative in the first place?

Bonobos and chimpanzees show us that flexible co-operation requires tolerance. Chimpanzees understood they needed to co-operate to get the food, but that was not enough. Their emotions got the best of them, and their co-operation fell apart.

It is a simple but powerful idea: Before humans could become ultra co-operative, we had to become ultra-tolerant. This tolerance preceded the evolution of more complex forms of human social cognition. Inferential reasoning, planning, and skills for co-ordination do little good in planning for hunting or finding shelter if no one can tolerate engaging in group activities or even listening to what others have to say. The brain enabling these sophisticated behaviours is very metabolically expensive and could not evolve if it were not immediately contributing to the individual's survival. This means it can only be after humans became more tolerant that unique forms of co-operative cognition could have evolved.

Selection against aggression made possible the ability to co-operate and communicate in foxes, dogs, and bonobos. Perhaps it did the same thing for humans.

Even beyond what we have observed in foxes, dogs, and bonobos, self-domestication may have also catalysed an evolutionary chain reaction leading to the evolution of completely new cognitive abilities – not just the expression of old cognitive skills in new contexts.

As a population shifted to become more tolerant and friendly, anyone with better social skills would have had an advantage over those who were less skilled. These more socially savvy individuals could actively seek out ways to benefit from their new capability of sharing the spoils of joint effort. For example, they may have begun sharing food with non-relatives, taking turns looking after one another's children, and even forming alliances with tolerant neighbouring groups to outcompete smaller, less tolerant groups. These socially sophisticated individuals would survive and reproduce more successfully generation after generation. Eventually, you have an entire population of highly tolerant and socially savvy individuals. Friendliness allowed humans to get cleverer.

Self-domestication could have facilitated an explosion of new forms of co-operation and communication that radically changed the course of human history.

Working Dog Intelligence

Self-domestication might help explain how social cognition evolved in humans even after an initial explosion of new forms of co-operation. As exciting as this idea is, it is difficult to test. Ideally, we would compare our behaviour with that of our distant human ancestors. If self-domestication led to new forms of cognition, we should be less aggressive and more tolerant, co-operative, and communicative than our ancestors. The problem is that our ancestors are extinct.

Once again, dogs come to the rescue. After the first dogs were friendly enough to interact with humans, the most socially sophisticated individuals in a population would have often fared best. This is because humans would probably have realized how useful these dogs were for hunting and herding. After many generations of the friendliest and most socially skilled individuals being the most likely to reproduce, breeds of dogs may have evolved more sophisticated cognitive abilities. We can test this by comparing working dogs with other breeds. If working dogs have more sophisticated cognition, a similar process may have led to the evolution of cognition in humans.

If working breeds were selected for their co-operative and communicative abilities, they should be better at following human gestures than non-working breeds. Victoria Wobber and I chose huskies and shepherds as our working breeds, since both respond to verbal and visual signals when transporting humans or herding animals. As our non-working breeds, our control case, we chose basenjis. From my introduction to them in Congo, I knew that while basenjis help their owners hunt, they are more like sight hounds – they simply chase and corner their prey and do not rely on human signals to hunt (similar to hunting behaviour observed in the dogs of the Mayangna people of Nicaragua). As our other non-working breed, we chose toy poodles, since they were largely bred based on their appearance.

The results were that while all four breeds were skilled at using human gestures, huskies and shepherds were more so than basenjis and toy poodles. Working dogs were three times more successful than non-working dogs at using multiple types of human gestures to find hidden food. It seems that while all breeds (including New Guinea Singing Dogs and dingoes) can use human social gestures, working dogs are the most skilled of all.

All dogs are skilled at using human gestures as a result of selection against aggression, but a second wave of selection acted on certain breeds of dogs that need to co-operate and communicate with humans in especially flexible ways. Humans may have intentionally bred together dogs who were the most socially skilled. Consequently, working dogs are even more skilled than non-working dogs are at using human gestures.

Our species's unmatched cognitive abilities may be the result of a related process. Self-domestication in humans created an opportunity for a second wave of selection to act directly on individual differences in social cognition. Those individuals with early-emerging social skills developed the most sophisticated cognitive skills as adults. They were more successful than other humans, since they could maximally benefit from the high levels of tolerance that evolved among group members. After many generations of these cognitively sophisticated individuals outcompeting everyone else, only their offspring remained. An extremely tolerant and cognitively advanced species evolved as a result.

The Road to Our Domestication

Even though our human ancestors are extinct, we can look for signs of self-domestication in the fossil record. Archaeologists have noted that morphological changes have occurred in humans that are consistent with self-domestication. In comparison with archaic *Homo*, who lived around 200,000 to 300,000 years ago, modern humans have thinner bones, smaller and more crowded teeth, a shortened face (creating our chin), and – perhaps most surprisingly – smaller craniums. Some data even suggest that the human brain has actually shrunk between 10 and 30 percent over the past 50,000 years.

This anatomical signal is present precisely when humans began showing modern forms of human behaviour. For example, in the last 50,000 years, complex cultural artefacts begin to appear in the fossil record, such as burials (with dogs), art, fishhooks, fire, shelters, and compound weapons such as arrowheads. Clearly, humans had begun co-operating in extremely complex ways that would equal our abilities even today.

Studies of modern hunter-gatherers point to a potential mechanism that may have been responsible for generating selection against aggression during human evolution. For the majority of our species's short evolutionary history we have lived as hunter-gatherers. Agriculture only began appearing in the Levant and China just over 12,000 years ago. Our current industrialized lifestyle is only a few generations old. This means all humans lived as hunter-gatherers for hundreds of thousands of years during the evolution of our species.

Our social system today is not representative of the social system of modern populations still living as hunter-gatherers. Research on these populations over the past century has revealed that they do not have a leader or dominant individual who makes decisions for the group. Instead, bands of hunter-gatherers work together against any individual who tries to dominate the group. Female bonobos co-operate against bullies; hunter-gatherers work together to ostracize, reject, and potentially even kill anyone who tries to forcibly rule the group. According to this cross-cultural research, human groups effectively work together to select against individuals who are overly aggressive towards members within their own groups.

This selection against the most aggressive individuals and selection for the friendliest social partners could have been enough to lead to a form of self-domestication late in our species's evolution. It could also be what has made us tolerant enough to adopt a wide range of urban lifestyles – many of which require extreme levels of tolerance. Think of the co-operation and tolerance required for three hundred people to board a plane, or for hundreds of drivers on a crowded city road. This tolerance allows for not only high-density living but also large population sizes – both factors related to the technological productivity observed across cultures. Larger populations produce more innovations that lead to more complex technologies.

The mostly peaceful high-density living we enjoy may be the result of self-domestication facilitating today's urban populations, which are so innovative.

Did Dogs Domesticate Us?

A more radical proposal suggests that our relationship with dogs explains the anatomical signature of self-domestication in modern humans. Colin Groves of the Australian National University has proposed that it was dogs who domesticated us:

> Dogs acted as humans' alarm systems, trackers and hunting aides, garbage disposal facilities, hot water bottles, and children's guardians and playmates. Humans provided dogs with food and security. . . . Humans domesticated dogs, and dogs domesticated humans.

During my trip to see the bonobos in Congo, Claudine and I found ourselves waiting patiently beside a small hut watching the embers of the fire die down from the previous night. We had travelled deep into the Lopori River basin in search of wild bonobos who might be living adjacent to Etumbe's group at the release site in Basankusu. But Claudine also had another goal. She wanted to help the friends we made on the plane find their basenjis, so she was investigating if anyone in the surrounding villages was breeding ba senjis. Finally, a tall, skinny man slid out of the forest with two relatively small dogs not far behind him. They had sharp, pointed ears and tight, curly tails like a pig's. They were not sure about Claudine and me, so they approached with caution. As the dogs came forward, we could see their shiny, tan coats, and we knew we had found what we were looking for. Claudine had met this man on a

119

previous excursion into the jungle, and he welcomed us like old friends.

The man looked tired. It was hard to find food in a forest that was already overhunted. I asked him how he managed to feed his dogs, given that it must be hard enough to feed his children. The man replied that he would rather sell his canoe or lose a child than let his dogs starve. He was completely dependent on his dogs to help him hunt what little prey there was to feed his family of ten. Without his dogs, he feared they all might starve.

As proto-dogs began relying on human refuse, each generation became increasingly friendly towards humans. Like ravens following wolves on a hunt, dogs probably began following human hunters. With keen instincts for detecting and chasing prey, dogs would have become a valuable advantage.

For instance, hunters in lowland Nicaragua rely on dogs to detect prey. The dogs run freely and bark when they have cornered an animal. The hunters recognize and follow the barks, until they locate and kill the prey. Hunters without dogs are usually less successful. Similarly, moose hunters in alpine regions obtained 56 percent more prey when accompanied by dogs.

Humans who were tolerant of proto-dogs would have been better off than those who were not. More tolerant and less aggressive humans could maximize the benefit of both a new night watchman who made it harder for enemies to raid camps successfully and a new hunting partner who led to more reliable supplies of meat. Increasing hunting efficiency gave humans more resources and allowed for more tolerance and sharing among group members.

Dogs could also be an emergency food supply. Thousands of years before refrigeration and with no crops to store, hunter-gatherers had no food reserves until the domestication of dogs. In tough times, the dogs who were the least efficient hunters might have been sacrificed to save the group or the best hunting dogs.

Although not a pleasant thought, harvesting dogs could have been one of the key innovations that eventually led to the invention of agriculture. Once humans realized the usefulness of keeping dogs as an emergency food supply, it was not a huge jump to realize that plants could be used in a similar way. It may seem far-fetched, but it is possible that we have been as strongly affected by our relationship with dogs as they have been by their relationship with us. Dogs may have civilized us.

When I think back to Oreo and my parents' garage, the way that a dog shaped my life, it does not seem such an absurd possibility.

PART TWO

DOG 'SMARTS'

6

DOG SPEAK

Are we having a conversation?

Since those first studies with Oreo in my family's garage, the field of dog cognition, or 'dognition', has exploded. From being thought of as an unremarkable animal made stupid by domestication, all of a sudden dogs have become one of the most popular species for animal researchers to study. More than a dozen research groups all over the world have turned to dogs to figure out how animal minds, including our own, work.

A lot of the initial research on dognition has focused on communicative abilities. We have seen that dogs are geniuses in their ability to read our gestures. Their skills are similar to what we observe in infants. The mental flexibility of dogs has led other researchers and me to suggest that dogs have a basic appreciation of our communicative intentions. They often use our behaviour to infer what we want, and what we want is usually to help them.

Communication is not just visual and does not just involve receiving and interpreting information, however. Communication can also be vocal and require producing meaningful signals. Do dogs understand words in the same way we do, do their vocalizations actually mean anything, and do they communicate differently depending on their audience? Researchers have conducted dozens of studies aimed at answering these questions in their search for a more nuanced understanding of dog genius.

Do Dogs Understand Our Words?

One morning, my dad looked at the newspaper at the bottom of our long, steep drive and decided that Oreo ought to start fetching it. Oreo was seven at the time, and I remember thinking that he was too old to learn new tricks.

Dad went to the bottom of the drive, pointed to the newspaper, and said, 'Fetch paper!' The Sunday paper was so heavy that Oreo's head sagged with the weight, but he wagged his tail when Dad patted his head and said, 'Good boy!'

Dad repeated this exercise each morning for a week. By the following Sunday, Dad just stood at the top of the drive and said 'Fetch paper!' and Oreo ran to get it. And just like that, Dad had a dog who could fetch the paper. I remember being amazed that it had not taken a hundred mornings to teach Oreo to retrieve the paper and wondering how he had learnt the word so quickly.

To answer this question, we need to revisit the story of Rico, who could remember hundreds of names for toys, and Chaser, who could remember more than a thousand. These dogs discussed in Chapter 1 were found to be capable of learning words through a process of exclusion. When Rico or Chaser heard a new word, each inferred that it was for a new toy. So they brought back a toy that they had not yet learnt a name for. Both dogs could remember the name–object pairings for a minimum of ten minutes after only hearing the new sound twice. Rico could remember some new names a month later. Amazingly, with very little practice, Chaser remembered all the words she was taught – and her vocabulary just kept growing.

Rico and Chaser's ability to use exclusion to link new human-made sounds to new objects and to remember so many of these names for so long is the closest thing researchers have seen to how children learn words.

However, when children learn a word like *sock,* they do not just associate the noise *sock* with one object that happens to be a sock. Instead, they understand that the name *sock* refers to all objects that function as clothing for our feet. The word *sock* represents a category of objects that can be different colours, shapes, textures, and sizes. The original study with Rico did not address the question of whether he understood that words refer to categories of objects. This led developmental psychologist Paul Bloom at Yale University to suggest that it is likely that 'babies learn words and dogs do not'.

Chaser's owner, John Pilley, designed a study with Chaser to address this concern. Pilley used different categories of toys: Frisbees, balls, and random dog toys of different sizes and shapes.

After Chaser knew the names of hundreds of toys from the different categories, Pilley introduced a new test, where he set out toys that were all from the same category – either Frisbee or ball. He then asked Chaser to fetch the toys by saying 'Fetch the . . .' followed by the category.

Because all of the toys belonged to the same category, Chaser was rewarded with praise regardless of which toy she retrieved. After she did this three times with a group of toys from each category, Pilley mixed toys from the different categories that had not been used in the introduction. He then asked Chaser to either fetch a Frisbee or a ball. Even though Chaser had never been tested with the toys this way, she always brought back the toy from the category Pilley had requested.

As a bigger challenge, Pilley did the same thing for a set of objects Chaser was familiar with from home, like shoes and books. Pilley put out some of these objects that had never functioned as toys, along with a random set of her toys, and asked Chaser to retrieve either a 'toy' or a 'non-toy'. Again, Chaser never made a mistake. It seems that while Chaser was learning the names of each new toy, like a human child, she also spontaneously placed each toy into categories (toy and non-toy) as well as subcategories (Frisbee, ball, and other).

As impressive as this might seem, a developmental psychologist would still be doubtful that dogs are learning words. This is because children also understand the symbolic nature of words. One of the simplest demonstrations of children's understanding of symbols is their ability to see the link between an object and a visual replica of the object. For example, if an experimenter hides a toy inside a triangular container, then shows a child a small replica of the triangular container, the child immediately understands to search inside the triangular container for the toy instead of other containers of different shapes. Children also learn that two-dimensional pictures can represent three-dimensional objects. They know that a toy that is one centimetre in a photo is much larger in real life.

We often use these layers of meaning when we communicate. Cuneiform was among the first written languages and consists of a series of pictographs. Computer icons are just another example of a symbolic system. Anyone who wants to print a document looks for the tiny replica of a printer on the toolbar. The designer employed an icon as an effective, succinct method of communicating the act of printing.

Juliane Kaminski, who did the original Rico studies, wanted to know if dogs could spontaneously use visual replicas as symbols. She designed a test similar to the one used with children. As in previous tests of Rico, Kaminski placed toys in a different room and asked Rico and several other Border collies to retrieve the toys one by one.

This time, Kaminski did not request the toys using words. Instead, she showed the dogs replicas of the toys and asked them to fetch. If the dogs understood the communicative nature of her request, they should retrieve the correct toys based on the visual representation.

When I first heard Kaminski was conducting this experiment, I thought there was no way a dog would spontaneously understand the link between a symbol and a toy. Happily, I was wrong.

All of the dogs spontaneously retrieved a toy that was represented by the replica. If Kaminski showed them a hot-dog toy, they went into the next room and brought back the matching hot-dog toy. Most of the dogs did this whether the replica was the same size as the toy or a miniature version of it. Rico and one other dog even retrieved the correct toy on their first trial when they were only shown a photograph of the toy. This spontaneous performance could only be possible if the dogs combined their understanding of our communicative intentions with an understanding of the symbolic nature of our helpful behaviour.

The latest experiments show that at least some dogs understand the categorical and symbolic nature of human communicative signals. Rico, Chaser, and several other dogs clearly show that at least some dogs use a variety of communicative skills that continue to match what we see in infants. No other species besides humans has demonstrated the ability to learn the meaning of words so quickly and with so much flexibility.

Every dog lover, including me, would love to know if these skills are somehow special to a handful of dogs or if most dogs have them. Since this research has only been conducted with Border collies, maybe these skills are found in only this breed.

But maybe Rico and Chaser are just the tip of the iceberg. The human parents of Rico and Chaser do not believe their dogs are unusually gifted. Neither dog was specially selected from a large pool of dogs who failed to show the same skills. Chaser, for instance, was randomly selected from a litter for the purpose of training her to participate in the studies. It would be quite a coincidence if the very first dog chosen just happened to break the world record for word learning.

A number of studies point to the possibility that a variety of dogs can make the type of inferences Rico and Chaser are using while 'fast mapping'. Remember that dogs can make inferences based on exclusion when searching for a toy or when judging new pictures on a computer screen. Dogs also fail a complicated toy-finding game but can spontaneously infer the solution when an experimenter communicates it.

There is also evidence that dogs are able to learn the names for new objects by simply hearing a conversation between two people. Two people spoke about a new object repeatedly before asking a dog to retrieve it using the new name and rewarded the dog only with praise, rather than food. While these dogs did not learn nearly as fast as Rico or Chaser, they did learn the names of two new objects

Our dog Tassie as a puppy. It was suspected that dogs are more socially skilled than primates and wolves because they are heavily exposed to people throughout their lifetime. In a surprise finding, young puppies with little exposure to humans are as skilled at using human gestures as adult dogs are. BRIAN HARE AND VANESSA WOODS

Rico was a Border collie discovered by Dr Juliane Kaminski. She demonstrated that Rico learned hundreds of words using inferential reasoning similar to that of a human infant. SUSANNE BAUS

Oreo was Brian's childhood dog. It was his ability to use Brian's pointing gestures while playing fetch that led to a series of scientific studies that revealed the social genius of dogs. BRIAN HARE

Christina Williamson collaborated with Brian to compare the social skills of wolves with dogs. Even wolves, like this one, who are highly socialized with people do not spontaneously use human communicative gestures like dogs. SHERMAN MORSS JR

Dmitri Belyaev risked his life in Stalin's Soviet Union to domesticate silver foxes experimentally. Brian's collaborator, Anna Stepika, holds the product of forty-five years of selecting foxes to be friendly towards humans. BRIAN HARE

Mystique the dog with Masisi the bonobo. Mystique guards the bonobo orphanage where Brian and Vanessa study bonobo behaviour and cognition. Bonobos are thought to be 'the dog of the ape family' because they show many characteristics in common with domestic animals like dogs. BRIAN HARE

A wild dingo on Fraser Island, off the east coast of Australia. Dingoes are like living fossils. They are probably similar to the very first dogs that evolved as a result of bold and friendly wolves taking advantage of rubbish created by the first human settlements. DR BRADLEY SMITH

New Guinea Singing Dog puppies. New Guinea Singing Dogs are very close relatives of dingoes. Brian tested a group of New Guinea Singing Dogs and found that even though they have not been bred to communicate with humans, they are still skilled at using human gestures. THE NEW GUINEA SINGING DOG CONSERVATION SOCIETY

A Mayangna hunter uses his dog to flush out prey. In Nicaragua, a number of tribes are depending on their dogs to help them chase and corner prey in the rain forest. Skilled hunting dogs are highly prized. MENUKA SCETBON-DIDI

The Hadza foragers with their new dogs. The Hadza are one of the last remaining hunter-gatherer groups that lives much like all humans lived for the majority of our species' evolution. Just as hunter-gatherers who encountered the first dogs would have recognized their potential as a hunting partner, some Hadza have occasionally begun using dogs to track injured prey. FRANK W. MARLOWE

A dog being tested at the Duke Canine Cognition Center. The centre is the first on-campus research facility dedicated to unlocking the secrets of the dog's mind. Pet owners are invited to bring their dogs to the centre to play fun problem-solving games which require dogs to make choices. By observing the pattern of choices the dogs make, conclusions can be made about how they solve (or do not solve) different cognitive problems. BRIAN HARE

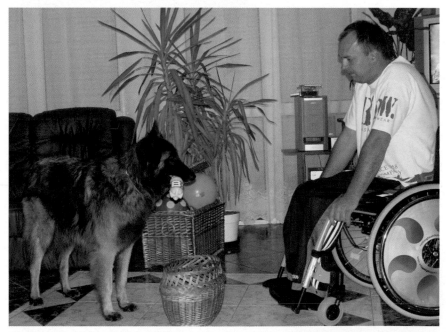

Phillip, the service dog tested by Dr József Topál. Dr Topál has demonstrated that Phillip is able to spontaneously imitate novel actions that he observes a human to make and then infer what a human can or cannot see. Both skills require a level of social complexity not previously attributed to dogs. JÓZSEF TOPÁL

Chocolate jealously looks on as Cina victoriously chews her ball after a game of fetch in the surf. Chocolate typically loses to Cina because Chocolate takes a direct route to the ball through the surf while Cina takes a shortcut down the beach before swimming the shortest distance possible to the ball. Most dogs are not particularly skilled at navigating or understanding simple physics. BRIAN HARE

Milo (far right), at the dog park in Germany, sniffs a stranger 'up close and personal'. Wolves have a different social system than dogs and do not tolerate these types of intrusions by strangers. Brian adopted Milo from a rehoming centre to live with him in Germany. Milo looked like a beautiful, fluffy Labrador but quickly showed he was mixed with chow chow. Brian learned that significant breed differences exist in behaviour and that Milo was as remarkable as Oreo. BRIAN HARE

Adults rate the same person as more attractive and trustworthy when photographed with a dog than when photographed without. BRIAN HARE

Breed-specific laws based on appearance as opposed to bad behaviour are doomed to fail in protecting the public because it is difficult to judge a dog by her cover. What breeds contributed to the genetic make-up of this dog? Because of his facial markings, most people think this dog is the offspring of the female Rottweiler he shares a back garden with. He is actually the son of Mystique and another village dog – both of whom look nothing like Rottweilers. BRIAN HARE as quickly by overhearing their names as when they were explicitly trained with more traditional food-rewarding techniques. It is likely, then, that most dogs learn to respond to a number of our words without any active training on our part.

Your dog may not be up to Rico and Chaser's level – or they might be above it. In any case, our pet dogs probably have some, if not all, of Rico and Chaser's inferential skills. We will know more as researchers continue studying the communicative skills of pet dogs.

As remarkable as dogs are at understanding us, this is just one side of communication. Communication does not just involve receiving information. Does the conversation go both ways?

Can Dogs Talk?

Mystique is a dog who lives at Lola ya Bonobo, where Vanessa and I study bonobos. During the day, she is sweet and demure, but at night she becomes a different animal. She guards our house, barking ferociously every time someone comes within earshot. Usually in Congo, a little extra security is appreciated. The only problem is that our house is on the main trail where the night staff walk back and forth after dark. Mystique dutifully barks at all passersby whether she has known them for a day or all her life. Eventually, we just learned to sleep through it. But if there was really a cause for concern, like a strange man with a gun, I wonder if Mystique would bark in a way that would alert me that there was something dangerous and different about the person approaching the house.

Dog vocalizations may not sound very sophisticated. Raymond Coppinger pointed out that most dog vocalizations consist of barking, and that barking seems to occur indiscriminately. Coppinger reported on a dog whose duty was to guard free-ranging livestock. The dog barked continuously for seven hours, even though no other dogs were within miles. If barking is communicative, dogs would not bark when no one could hear them. It seemed to Coppinger that the dog was simply relieving some inner state of arousal. The arousal model is that dogs do not have much control over their barking. They are not taking into account their audience, and their barks carry little information other than the emotional state of the barking dog.

Perhaps barking is another by-product of domestication. Unlike dogs, wolves rarely bark. Barks make up as little as 3 percent of wolf vocalizations. Meanwhile, the experimental foxes in Russia bark when they see people, while the control foxes do not. Frequent barking when aroused is probably another consequence of selecting against aggression.

More recent research indicates that there might be more to barking than we first thought, however. Dogs have fairly plastic vocal cords, or a 'modifiable vocal tract'. Dogs might be able to alter their voices subtly to produce a wide variety of different sounds that could have different meanings. Dogs might even be altering their voices in ways that are clear to other dogs but not to humans. When scientists have taken spectrograms, or pictures, of dog barks, it turns out that not all barks are the same – even from the same dog. Depending on the context, a dog's barks can vary in timing, pitch, and amplitude. Perhaps they have different meanings.

I know two Australian dogs, Chocolate and Cina, who love to play fetch on the beach. Each throw sends them plunging through the waves, racing for that magic orb of rubber. When Chocolate retrieves the ball, inevitably Cina wrestles the ball from Chocolate's mouth, even while Chocolate growls loudly. The girls also eat together, but when Cina tries the same trick with Chocolate's food, the result is very different. A quiet growl from Chocolate warns Cina away.

It is difficult to see how Cina knows when it is okay to take something from Chocolate's mouth, since both growls are made when Chocolate is aggravated and unwilling to share. If anything, Chocolate's growl seems louder and scarier when she is playing than when she is eating.

Experiments have now shown that dogs use different barks and growls to communicate different things. In one experiment, researchers recorded a 'food growl' where a dog was growling over food, and a 'stranger growl' where a dog was growling at the approach of a stranger. The researchers played these different growls to a dog who was approaching a juicy bone. The dogs were more hesitant to approach if they heard the food growl rather than the stranger growl.

In another experiment, researchers recorded 'alone barks' of dogs when they were alone, and 'stranger barks' when a stranger was approaching. When researchers played three 'alone barks' to different dogs, these dogs showed less attention to each bark. But when they played the fourth bark, the 'stranger bark', the dogs quickly jumped to attention. They did the same thing when the barks were reversed, showing that dogs could clearly distinguish between the two types of barks. Using a similar test, the dogs also distinguished between the barks of different dogs.

How well do people understand what dogs are saying? Researchers played a collection of barks to a group of people. Regardless of whether they owned a dog or not, most people could tell from a bark whether a dog was alone or being approached by a stranger, playing or being aggressive. Unlike dogs, people were not very skilled at discriminating between different dogs. The only time people could tell between different dogs was when they heard the 'stranger bark'. This is the exact moment a dog owner would be most likely to want to understand the meaning of a dog bark, since strangers can mean trouble.

These initial studies show that growls and barks do carry meaning that other dogs and, in some cases, people can recognize. This complexity comes as a surprise. Of course, our dogs have known all along – just ask Chocolate and Cina. Still, we know very little about the vocal behaviour of dogs.

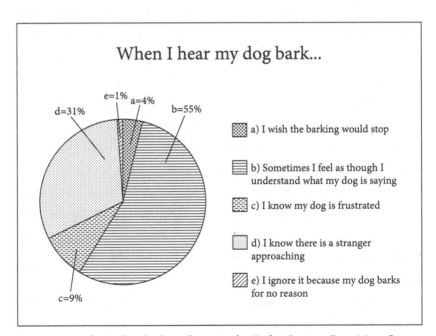

When I hear my dog bark...

e=1% a=4%
d=31%   b=55%

a) I wish the barking would stop

b) Sometimes I feel as though I understand what my dog is saying

c) I know my dog is frustrated

d) I know there is a stranger approaching

e) I ignore it because my dog barks for no reason

c=9%

In a survey of people who have been to the Duke Canine Cognition Center and taken an online survey at www.cultureofscience.com, 86 percent of people feel like they sometimes know what their dog is trying to communicate by barking.

Not only can we understand some of what dogs say, it seems dogs might understand whether or not we can hear them. My Australian canine friend Chocolate is crazy about plush toys. When a friend gave our daughter, Malou, a plush Santa Claus toy for Christmas, Chocolate was more excited than anyone. She became almost uncontrollable when she heard it squeak. We quickly realized there was serious concern for Santa's safety. We tried a few firm 'No' commands, and Chocolate reluctantly left Santa alone.

During dinner, we heard Santa's muffled squeak. Usually, it was Chocolate's great pleasure to de-stuff and de-squeak her dog toys in front of her humans. But this time, Chocolate had taken Santa into the back bedroom, as far away from us as she could possibly get. Luckily, Santa's squeak saved him, although Chocolate did manage to tear the pom-pom off his hat before we came to the rescue. Did Chocolate just want to be alone with her new toy, or did she take the humans into consideration, hiding in a room where she could not be seen or heard?

In one experiment, two open boxes were placed on their sides and strings of bells were attached to the openings. To get to the food inside the box, the dogs had to push through the bells. The trick was that one set of bells had their ringers removed.

Once the dogs were familiar with the noisy box and the quiet box, a human experimenter put food in both boxes and forbade the dogs to take it. The human then stood between the two boxes. In the condition where the human was facing the dog and the boxes, the dogs pilfered from either box, but when the person's back was turned, the dogs avoided the box with the functioning bells. Even more amazing, they did this on their very first trial.

So even though dogs might seem to make a racket regardless of who is around, there is some evidence that they are well aware of who is in range and what they can hear.

Do Dogs Use Gestures?

We do not just communicate with our voices. We also use visual signals. Entire languages, such as sign language, are based on visual gestures. Infants begin pointing around the same time they start to understand the gestures of others. While dogs do not use gestures as complicated as sign language, researchers have begun investigating whether dogs can produce visual signals to communicate in similar ways to infants.

Dogs frequently use visual signals to communicate with others during their natural interactions. When dogs are playing, careful observations have shown that gestures such as play bows (dropping the chest to the ground and preparing to spring back at the first sign of a chase) are used to signal that their behaviour is just meant in good fun. Where a quick approach and forceful contact might usually lead to aggression, the same behaviour preceded by a play bow leads to a friendly response. When other dogs see a play bow, they can agree to the game with another visual gesture. They usually do this using a 'self-handicapping' gesture that makes them more vulnerable (rolling on their back). This means dogs might flexibly use visual gestures to communicate, perhaps even more than vocal behaviour.

Experiments have shown that dogs use visual gestures to communicate intentionally what they want in new contexts. Once we realized that Oreo was so skilled at understanding what we were trying to tell him, we started to wonder how much he could tell us. So I strung up three wicker baskets between two trees in my parents' back garden. Then I asked my little brother Kevin to hide food in one of the three baskets while I was inside the house. Kevin would show Oreo which basket he put the food in and then walk away. With the baskets out of Oreo's reach, there was no way for him to get at the food.

When I came outside, I had no idea which basket had the food. My only hope for finding the food was to watch Oreo's behaviour. Without words or a finger to point with, Oreo was crystal clear. He ran underneath the basket with the food inside, alternated his gaze between me and the basket, and barked. If I did not come outside, Oreo did nothing. Since Oreo used communicative behaviours only in my presence, it suggested he was intentionally behaving this way to help me locate the food.

Subsequent research with larger samples of dogs has shown that Oreo was not unusual. Most dogs use similar 'showing' behaviour, and they can be very persistent at enlisting human help. But dogs do not use showing behaviour if a human appears but nothing has been

hidden. These studies show that dogs not only understand what we tell them, they can also tell us a thing or two.

Dogs do not have the right vocal cords to mimic human speech like a parrot, but one day they might be able to use more sophisticated language. Alexandre Pongrácz Rossi of the University of São Paulo in Brazil has already taken a step in this direction. In the first study of its kind, she trained her dog Sofia to use a keyboard with symbols that represented a walk, food, water, a toy, play, and her crate.

If Sofia pressed one of the symbols on the keyboard, the keyboard would say the word. Sofia became skilled at pushing the correct key in the correct context. She would run to the keyboard and push the symbol for toy when Rossi bought her a new toy. Sofia also would alternate her gaze between Rossi and the toy while pushing the key for toy, in what seemed like clear attempts to communicate. When Sofia licked her lips and ran to the keyboard, she always pushed the key for either water or food. Also, Sofia never used the keyboard when she was alone.

Given Sofia's skills at communicating with these symbols, will she be able to learn more symbols? Will she ever string two symbols together to say 'walk water' by way of communicating that she wants to go to the lake, or 'crate toy' when she wants a toy that is locked in her crate? Dogs will never show the astonishing skill of human infants as they start to talk and use gestures, and they may never match the great apes (Kanzi the bonobo communicated using a keyboard with 348 symbols); but future research could help us understand when and what dogs want. It may also help us understand more about how they view their world. For instance, when Sofia was introduced to her first guinea pig, she quickly ran to her keyboard and kept pressing the key for food (as opposed to toy!).

Do Dogs Adjust to Their Audience?

Dogs can produce at least a few vocal and visual signals that humans and other dogs can understand. This means they might be intentionally trying to communicate with us. If this is true, then dogs should not just reflexively make signals when encountering set environmental stimuli, like barking at a stranger. They should be communicating about their environment to others, like barking to recruit others to help warn off a stranger. One way to determine if animals are intentionally communicating is to see if they adjust their signalling depending on whether their audience will receive their signal.

Olfactory signals, or smells, are used by a variety of animals, including insects. Olfactory signals are probably used so widely because they do not require intentional communication. Strong smells are inescapable, they linger, and they are easy to place in many locations. This means animals do not have to understand much about their audience; eventually their signal will be received.

Visual signals are far more subtle and personal. They are only briefly available and can only be received if the intended audience is watching. If signallers visually communicate while their audience is not present or looking away, their signals can easily be missed. Research is beginning to show that dogs are sensitive to what their audience can and cannot see. This supports the idea that dogs intentionally communicate with us as well as with other dogs.

When I was young, whenever I was outside, Oreo would bring me a tennis ball. Regardless of whether I was raking leaves, talking to a friend, or anything else that clearly indicated I was busy, Oreo would drop a ball at my feet. Sometimes I tried to ignore him by turning my back or gazing off into the distance. But Oreo would make sure he dropped the ball where I could see it. Oreo's appreciation of my visual field was literally inescapable.

Inspired by Oreo, we conducted a simple experiment. I would throw a ball for Oreo and before he returned, I would either face him or turn my back to him. Oreo always dropped the ball in front of me where I could see it, and on the few trials where Oreo did not, he nudged the ball into my back and began barking.

Alexandra Horowitz from Barnard College in New York conducted a study to examine how dogs communicate with one another in their natural interactions. Horowitz videotaped hundreds of hours in a San Francisco dog park, focusing on moments when one dog was trying to initiate a social interaction with another dog, like a play bow. Based on her observations, she wanted to know if dogs were using visual signals to initiate social interactions when the intended recipient could see them.

Horowitz slowed down the footage and found that in most cases, dogs were using visual signals, like play bows, only when their target was facing them and could see the signal. Horowitz also saw that when their target was not facing them, rather than using visual signals, dogs were more likely to use a tactile behaviour, like touching the other dog with a paw.

An experiment showed that guide dogs who live with a blind person are more likely to make a sound, like licking with their tongue, when indicating the location of food than dogs who have not been trained to work with blind people. The licking behaviour is not a part of the dogs' training but is a result of living with a person who does not respond to visual signals. Systematic observations of natural behaviour suggest that dogs might adjust their communication based on what others can and cannot see.

In another experiment, a dog could choose to beg for food from one of two people. The trick was that one person could not see them. For example, one person was wearing a blindfold over their eyes while the other person was wearing a blindfold over their mouth. Several research groups have tested dogs in this type of situation and have found that dogs prefer to beg from a person who is facing them or has their eyes open rather than someone wearing a blindfold or sunglasses. These studies show that dogs are sensitive to subtle cues of our attention. Dogs know that if they can see your face and eyes, they will be able to communicate with you.

Based on results like these, further experiments explore whether dogs are sensitive to our visual perspective. We take the visual perspective of others all the time. If someone asks to borrow a pen, I can infer they are referring to the one they can see in my hand, even though I can also see a pen on the floor behind them. Taking someone else's perspective helps me figure out the answer when it is not in front of me, by using someone's eyes or face. Visual perspective taking requires me to abandon an egocentric view of the world and think about what others can and cannot see when they communicate.

Juliane Kaminski and her colleagues designed an experiment where someone placed one identical ball behind each of two barriers. One barrier was transparent, and the other was opaque. The dog was placed on the side of the barriers that allowed her to see both balls. A person was on the opposite side and could see only one ball through the transparent barrier. Then the person asked the dog to 'fetch' and rewarded her with some playtime, regardless of which ball she retrieved first.

Dogs were more likely to fetch the ball that the person could see, even though there was no reward for doing so. In a control condition where the person was on the same side of the barriers as the dog, the dogs just brought back the balls randomly.

At least in this situation, dogs were responding to the communicative request of the person based on what the person could and could not see. Dogs may monitor what we can see when they are listening or trying to communicate with us.

Rather than just reading the behaviour of their audience, perhaps dogs are adjusting their communicative strategies to what their audience knows and does not know. One of the earliest ways that infants show an appreciation for someone else's knowledge is by tracking what someone else has seen in the past. For instance, if an infant sees an adult searching for an object, the infant will help the adult locate it. If the adult was out of the room when someone entered the room and hid the object, the infant will show the adult where the object was hidden. But if the adult was present when someone else hid the object, the infant is much less likely to point out the object. Infants as young as twelve months inform adults based on what the adults have seen in the past.

For a long time, researchers have thought that we are the only species that can recognize when someone is knowledgeable or ignorant, but now they have begun to challenge this accepted wisdom. Phillip is a Belgian shepherd (also known as a tervueren) who was trained as a service dog. József Topál from Eötvös Loránd University in Hungary decided to conduct some initial tests with Phillip to test whether dogs can discriminate between someone who is ignorant and someone who is knowledgeable. As a service dog, Phillip was trained to indicate the location of objects as well as retrieve them for his owner.

Topál set up several boxes in which he could hide various objects. Each box could be locked, and the keys could be hidden. The question was whether Phillip would know when to communicate the location of the keys and the hidden object based on what someone had seen. Phillip might always help someone locate the objects, or like an infant, he might adjust his efforts depending on what someone had seen or not seen in the past.

Phillip's behaviour turned out to be similar to that of infants. If the person did not see Topál hide the object, Phillip would fetch the keys and orient towards the correct hiding location. But if the person had been present while Topál hid the toy, Phillip was much less likely to help point out where either the key or the object was hidden.

Buoyed by the initial results with Phillip, Topál ran a larger study with a group of dogs to see if they could replicate these findings. He found that even pet dogs might be sensitive to what others have seen in the past. Dogs were more likely to indicate the location of a hidden toy with barking and head movement if the person had not seen the toy someone had hidden. The researchers themselves went on to argue that dogs might discriminate between people based on what they know or do not know.

In an equally rigorous study following the same method, this finding was not replicated. A related study also found that while dogs intentionally make requests with showing behaviour, they only do this for objects they want. A dog will not inform a human of where an object is hidden if the human is interested but the dog is not. This suggests dogs may not understand when others are knowledgeable or ignorant. It also suggests that when dogs 'speak' to us, it is less motivated by wanting to help us and more motivated by wanting us to help them.

Given the success of Phillip and other service dogs, perhaps the results depend more on the population of dogs tested and less on the potential for dogs to inform us.

The Conversation

Far from being one-sided, the conversation between dogs and people is quite sophisticated relative to what some scientists, including myself, have suspected. Dogs make different vocalizations with different meanings. Dogs and people can distinguish the contexts in which some dog vocalizations are made. Dogs can even recognize individuals based on these vocalizations. The communicative behaviours of dogs are not just uncontrollable noises they make when they become aroused. Dogs adjust their visual and vocal signals based on what will increase the likelihood that their audience will receive their signal.

Based on the current research, these communicative skills are similar to those observed in a number of other animals. Dogs do not even remotely approach the level of infants, who begin using gestures and words at the same time as they begin to acquire grammar. Also, while dogs can request things using visual signals, they do not use these same behaviours to inform others (except, perhaps, for Phillip).

Relative to other animals, it is the ability of dogs to understand human communication that is truly remarkable. Some dogs have the ability to learn hundreds of names for objects. They learn these names extremely rapidly through an inferential process of exclusion. They also spontaneously understand the category to which different objects belong. Some dogs even show understanding of the symbolic nature of human object labels. Dogs may truly understand words.

So far, we have focused on how a dog might be able to solve problems through communication. But dogs also solve problems when they cannot rely on other pack members, when there is no one around to communicate with. How does a dog do then?

7

LOST DOGS

Dogs don't beat wolves at everything

Even though humans are a super-social species, we often have to solve problems on our own. We cannot rely on others to help us back our car out of the garage, judge whether a box will fall off the shelf, or reflect on what we know and do not know. Our success as a species is not just a social story. As we grow, we rapidly become aware of the forces that shape the world around us – for instance, gravity. We also have to travel from point A to point B without getting lost.

The world of dogs is not so different.

Lost in Space

All animals must find food, water, and shelter in order to survive. It is also essential to be able to navigate through space and learn where things are. However, dogs (and domesticated animals in general) are a special case, since these survival problems have been solved for them by humans who provide all of the basic necessities (and sometimes a great deal more). Revisiting the notion of domestication making an animal dumber, some have wondered if domestication has led to the decline of the navigational abilities of domestic animals, since they never need to go farther than their front door.

Every morning when I am in Australia, I take my two dog friends Chocolate and Cina to the beach. On the way home, we have to go up a steep hill. Chocolate always trots dutifully up the hill with me, but Cina only goes up halfway, then cuts across the gardens of several neighbours, taking a shortcut home.

Dogs have recently been tested for their ability to take shortcuts. In the experiment, a dog watched as someone hid a piece of food in a large field. The experimenter then walked the dog up to thirty metres away from the food, took a ninety-degree turn, and headed ten metres in another direction.

At this point, the fastest route to the food was no longer the path they had just taken. Then the dog was released to find the food. This might sound like an easy test, except that after the food was hidden, the dog was blindfolded, the food was camouflaged, and the dog had its ears plugged. The dog could not see the food and, perhaps more importantly, could not smell it or use any environmental information to find it. Instead, the dog had to mentally calculate the shortest route back to the food rather than relying on sensory information.

In 97 percent of trials, the dogs found the food in just over twenty seconds on average. This means dogs can solve primary school trigonometry – they take the hypotenuse of the triangle, the shortest distance between their location and the reward.

On our morning walks, Chocolate and Cina chase balls through the waves. Chocolate is more of a city dog, while Cina grew up on the beach. Chocolate impulsively blasts through the water towards the ball regardless of where I throw it. Cina adjusts her strategy depending on where I throw the ball. If I throw the ball straight out, Cina is right next to Chocolate. But if I throw the ball parallel to the beach, Cina takes a detour. She runs down the beach until she can swim the shortest distance to retrieve the ball. Sometimes she even leaves the water, runs on the beach, and then plunges back into the water. This could be a flexible skill based on an understanding of detours, or it could be an inflexible trick she learned by chasing thousands of balls at the beach.

The evidence suggests that dogs might be a bit lost in space when it comes to barriers and detours. In one experiment, dogs had to go round a barrier to reach their owner, who rewarded them with food and praise. In the initial four trials, the dogs used an opening on one side of the barrier to reach their owner.

But then the opening was moved to the other side of the barrier – and none of the dogs could solve this simple detour problem on the first trial.

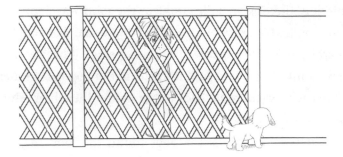

Even though the dogs could see the new opening and the old opening was no longer there, the dogs still looked for the old opening. It took several trials for dogs to figure out to head directly to the new opening – and some dogs never solved the problem at all. When it comes to finding their way, sometimes it is difficult for dogs to realize that a previously effective strategy is no longer viable. This could explain problems that dogs encounter during agility or service dog training.

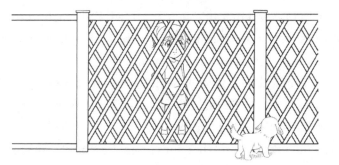

If dogs cannot even solve a simple detour problem, surely they would get lost in a maze. More than two hundred puppies of five different breeds were tested on their ability to find their way through a maze that required them to make ten different turns and avoid five dead ends.

And even though the maze was only 5.5 metres long, the puppies wandered round in it for two to seven minutes on their first trial before they eventually found the exit and received some fish and a big hug from an experimenter. The puppies did improve with practice. By the tenth trial, all of the puppies were exiting the maze in under one minute. Interestingly, different breeds used different strategies. Beagles got distracted by their need to sniff in every corner, fox terriers were slowed down by their earnest attempts to bite through the walls, and basenjis patiently but quickly made their way through the maze. Most puppies made about twenty errors in the first few trials, but a few talented puppies made almost no mistakes. They slowly weaved through the maze, carefully considering which direction to go next at each turn. This is quite remarkable, given that most puppies found themselves in around twenty dead ends in the first few trials.

But even the top performers did not really understand what they were doing. They soon developed a left–right strategy where no matter what they could see in front of them, they alternated between turning left and right. The more the talented puppies did the maze, the worse they got, until their performance resembled that of the less talented puppies.

It gets worse when you start comparing dogs with other species. Harry and Martha Frank, from the University of Michigan, tested how wolves and dogs solve simple detour tasks. Food was placed on the opposite side of a mesh fence that was either short (about a metre), long (about seven metres), or U-shaped. A group of wolves and dogs were tested for their ability to round the detour and obtain the food.

While wolves quickly solved all of the various detour problems, dogs constantly scratched at the food behind the fence or repeatedly reversed their direction before reaching the end of the fence. On the long fence, dogs made twice as many errors as wolves, while on the U-shaped fence they made almost ten times more errors. Remember that one criterion for genius is outperforming your close relatives. It looks like in the detour task, wolves get the genius award.

When comparing the memory of dogs and cats, there might be hope. In one experiment, a dog or a cat watched while an experimenter hid a reward in one of four boxes. Both dogs and cats had to wait before they were allowed to search for the reward. Cats quickly started to forget where the food was after only ten seconds, while dogs gradually started to forget where the food was after thirty seconds. By sixty seconds, cats almost completely forgot where the food was, yet even after four minutes, dogs still had some success at remembering.

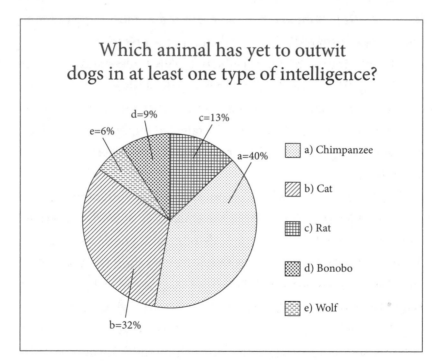

Although a third of people correctly guessed that cats have yet to outwit dogs, even more believed that chimpanzees cannot outwit a dog. In fact, chimpanzees can outwit dogs in most contexts – with the notable exception of communicating with humans.

But while dogs might have better memories than cats, that does not necessarily mean much. In a task used to test ageing and memory, dogs and rats had to go through a maze with eight arms radiating out from a central starting position such as a sun. Food was hidden in each arm, and the animals had to remember which arms they had already visited. While dogs were correct in 83 percent of their searches, rats were successful in more than 90 percent of their first searches. In a slightly more difficult version of the game, rats showed 95 percent accuracy in their choices while dogs were correct only 55 percent of the time.

Given all the news stories about dogs with SatNav-like capabilities, it might seem surprising that dogs cannot round a fence like a wolf or remember where they have been like a rat. Prince, a shaggy-haired shih tzu, found his way home after five years, even though his owner had moved four times. Mason the terrier was blown away in a tornado and came home despite two broken legs.

In 1991, Jacquie, my mother-in-law, drove with her dog, Snooper, to her friend's house, four miles from home. She briefly left on an errand, and Snooper busted out of the house and ran away. It was early evening in winter and pitch-black dark. After the search party was called off at midnight, Jacquie drove home. When she arrived, Snooper was sitting in the drive, wagging her tail. How did Snooper find her way home?

In some cases, dogs can use landmarks. Using landmarks makes it easier to find a target from multiple positions. Dogs watched an experimenter bury a toy under a uniform layer of wood chips that covered the entire floor. The toy was so effectively hidden that the dogs were unable to find or smell it. Then the experimenter placed a wooden peg within half a metre of where the toy was buried. Dogs were able to use the wooden peg as a landmark to find the toy.

Even when the experimenters shifted the peg when the dog was not looking, the dogs shifted their search proportionally to the new position of the peg. Dogs would not have shown this proportional shift if they were not using the landmark to guide their search.

But while dogs *can* use landmarks, usually they don't. Say a treat is hidden to their left. Then you turn the dogs around, so they are coming at the treat from the opposite direction. This means the treat will be on their right. Unfortunately, dogs still go to the left. This is called using an egocentric approach.

My mother-in-law's dog and all of the other dogs who find their way home might just be the lucky ones. Ádám Miklósi agrees when he says, 'Most lost dogs never find their homes.' Of the six to eight million dogs and cats who enter shelters every year, around 30 percent of the dogs are reclaimed by their owners. Many of these dogs were wandering about lost. Since dogs are fairly unremarkable in their ability to solve navigational problems involving detours, landmarks, or working memory, don't rely on your dogs finding their way home – have them microchipped instead.

Could a Dog Pass Physics 101?

Even as infants, humans quickly show an understanding of basic physical principles. At an early age, infants begin to understand that toys fall down if they are dropped, that a ball cannot pass through a solid object like a wall, and that when two toys are connected, if you move one, the other will move, too. This basic knowledge of physics helps us decide how to find, retrieve, and place things that we need.

When thinking about whether dogs understand the same principles, I always remember my walks on cold days in Germany with my dog Milo. When I adopted Milo, I had no idea it meant I would be walking five miles a day as a result. Most people ride bicycles around the city of Leipzig. This was out of the question with Milo, who had the unfortunate habit of passing any tall post on the opposite side from me. When he was wearing a lead, we would both end up tied round the post – a disaster if I was travelling twenty miles an hour on a bike. So until Milo learned to follow me while detouring around such posts, I decided to walk – which ended up being my permanent mode of transport. I suspect Milo was not the only dog with this problem. Even Oreo would quickly find himself wrapped around a tree if he was tied to one.

Experiments have shown that dogs do not understand the principle of connectivity. Harry and Martha Frank gave Alaskan malamutes and wolves a variety of tasks where they had to pull a rope to bring a food dish within their reach. Only the wolves immediately solved the various rope-pulling problems; the dogs never solved the more complicated versions.

Another experiment showed that it is difficult for dogs to learn to use a string to pull food out of a transparent box. At first, dogs ignored the string. They kept scratching at the food through the transparent box top; it took dozens of trials for them to discover the solution by accident.

Even after dogs learned to pull the string to pull the food out, if the position of the string was only slightly altered, they could no longer solve the problem.

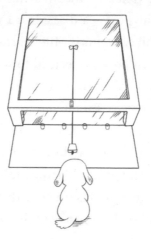

And if the food was moved close to the box's opening, they forgot about the string altogether and resorted to an ingenious but totally ineffective 'licking' technique of trying to stretch their tongues to grab the food.

Dogs did slowly get better, but another experiment showed that they were not learning anything about the cause of their success.

Instead of one string, there were two strings crossed in the shape of an X, only one of which was attached to the food. Dogs were attracted to and then pulled the string end closest to the food. They did not understand that the string needed to be connected to the food.

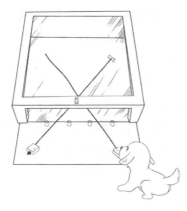

In comparison, both primates and ravens (the 'apes' of the bird world) tend to be skilled at solving a series of related problems (although some dogs may succeed in related tasks).

When it comes to understanding connectivity, dogs look as bad as cats, who also get tangled up on the same tests. This explains why I did not ride a bike in Leipzig and why dogs should not be left unattended when they are tied to a tree.

Dogs are famous for their keen sense of hearing. They often hear things that are out of humans' hearing range. They certainly understand the vocalizations of others, but they also hear non-social noises. If a dog stands at the edge of a waterfall, would she know that the crashing sound is the force of the water and that she should stay away from it?

A simple experiment suggests the dog might not. A group of dogs were given the choice of searching for food in one of two containers after a human shook one of the bins. Sometimes the container made a noise and sometimes it did not. If dogs understand that objects create noise when they collide, they should look in the container that makes a noise rather than the one that is silent. However, unlike chimpanzees, dogs always chose the cup a human touched, regardless of whether it made a noise when it was shaken or not.

A matter of physics that dogs might understand is the principle of solidity, which states that one object cannot pass through another. For instance, your dog may understand that you cannot throw a ball through a sofa or a house. To test this, dogs were shown two wooden boards. The experimenter picked up one of the boards and hid food underneath, creating an incline. The experimenter picked up the second board in the same way but left it lying flat. Since food cannot pass through the board, the food must be propping up the board. Dogs seemed to make an inference, because they had a strong preference for choosing the board on an incline.

In a related experiment, dogs were shown a piece of food sliding down a tube into a box. The food rolled to the far corner of the box and could be accessed only by the door farthest from the slide. Then a physical barrier was inserted, dividing the box into two. Now the food could be accessed only through the door closest to the slide.

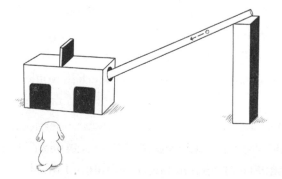

All of the dogs alternated their search for the food based on the presence of the barrier. The dogs seemed to infer that when the barrier was present, the food could not pass through it into the more distant part of the box.

However, even though dogs might understand solidity, they get a little confused when solidity is confounded with gravity. Dogs watched an experimenter drop food through a tube into one of three boxes. Sometimes the tube went straight down into a box, and sometimes the tube went into a box that was not directly underneath.

When the tube went straight down, dogs understood that gravity would make the food fall into the box underneath. But when the tube angled into a different box, dogs did not understand that the tube would prevent gravity from allowing the food to fall straight down. Unlike young infants and some primates, dogs were able to improve their performance with practice. This suggests dogs can overcome their gravity bias. The lack of spontaneous success makes it unlikely that the dogs understood what had caused the dropping food to defy gravity, however.

From what we know so far, it is unlikely that a dog will win a Nobel Prize in physics anytime soon. Connectivity leaves them bewildered, and although they seem to have a basic understanding of solidity, they get confused when gravity enters the picture.

Do Dogs Know Themselves?

In my very first experiment with Oreo, I tested his ability to follow my pointing gesture while he was swimming after tennis balls. I had seen that if he lost a ball in the lake, he would swim towards me until I pointed. This suggests a certain amount of self-knowledge – Oreo knew he did not know where the ball was and was asking for information from me as to where to look.

Another experiment that supports this possibility showed that dogs are as skilled as great apes at remembering what type of biscuit has been hidden in a jar. Like apes, if dogs see someone put food inside a container, they show signs of surprise if a different type of food is removed. They see you put a biscuit in the jar, and they expect the same biscuit to come out. Their surprise at seeing something different suggests they 'know' they know what is inside.

Another suggestion that dogs might recognize their own ignorance is that they quickly ask a human to help them when they encounter an impossible problem. When food is locked inside a container, dogs quickly look to a human and stop trying to open the box themselves. Perhaps this is not just because they get frustrated and always approach their parents when things get hard; instead, they know they do not know the solution and need help.

However, studies designed specifically to examine whether dogs know if they know something have found little evidence that dogs are aware of their own ignorance. If chimpanzees do not know where food is hidden, they will inspect different hiding locations before making a choice. For instance, if food is hidden inside one of two tubes, they will lean over to inspect inside the tubes to find the food. Yet, in two different experiments, dogs have failed to show the same type of self-reflective behaviour. In both studies, dogs immediately made a choice, even though they did not know where the food was. They just guessed randomly, despite being given the opportunity to inspect both hiding locations before choosing.

Understanding whether you have or have not seen something is just one type of self-knowledge. Another type is the ability to distinguish yourself from others. In the 1970s, Gordon Gallup developed the mirror test, in which an animal is placed in front of a mirror for several hours. While most animals behave as if there is a different animal looking back at them from the mirror (continuously threatening the mirror or looking behind the mirror for the other animal), a few species, including the great apes, quickly change their behaviour. They stop behaving as if they are looking at another individual and begin to use the mirror as a tool to see parts of their own body they normally cannot see. Bonobos will count their teeth, chimpanzees will straighten their eyebrows, gorillas will inspect their silver backs, and orangutans will peer behind their giant cheeks.

Dogs are among the species that show no sign of understanding their mirror image. After barking at the mirror or searching behind it, they tend to lose interest quickly.

Marc Bekoff, one of the best-known dog experts from the University of Colorado, was sceptical of the mirror test, however. He started watching his dog Jethro sniff other dogs' urine in the fresh Rocky Mountain snow. Jethro was very discriminating where he sniffed and where he marked with his urine. Bekoff saw that Jethro spent very little time re-marking snow that he had previously marked. Instead, he spent more time inspecting and marking where other dogs had marked since he had been on his last walk. Jethro could clearly discriminate his advertisements from those of other dogs.

Most dogs are also capable of assessing the size of other dogs relative to a sense of their own size (although this ability might be missing in a few spirited Parson Russell terriers). So while dogs know a lot about how to understand others, perhaps they have a more limited ability to reflect on themselves. They are able to discriminate themselves perceptually from other dogs (through scent, for instance), but we have little experimental evidence suggesting that they have a sense of self like we do, or that they can reflect upon what they know and do not know.

Going Solo

One of the most impressive things I have ever seen was a Chihuahua doing a complicated series of ballet manoeuvres to Justin Bieber's pop-chart hit 'Baby'. The sport is called 'musical canine freestyle', and the Chihuahua had learnt an amazing combination of spinning, twisting, and bowing that could rival Biebz himself.

Because you can teach dogs many fancy tricks, you might think they are exceptional at associative learning. Yet, even in associative learning, dogs are still unimpressive. Harry Frank thought that dogs would be faster than wolves in learning to associate a coloured block with the location of food. A white block indicated where food was hidden, while two black blocks were used to mark empty locations.

Frank reared a group of wolf puppies by hand and compared them with a group of malamutes. While the wolves took just over fifty trials to master choosing the white block as opposed to the black blocks, dogs required more than a hundred. When Frank switched the association so that the black block indicated the location of the food and white blocks did not, it took the dogs 30 percent longer than the wolves to master the task. When it comes to going solo, dogs should not go up against a lone wolf.

Victoria Wobber and I conducted a similar study, but in addition to an arbitrary association (the colour of the correct hiding location), we also ran a social version of the test. We thought perhaps dogs performed badly against wolves because they were learning something arbitrary. We predicted that if dogs had to use a more familiar clue, such as the identity of a person, they would do better.

In our social test, one person always rewarded the dog while another person never rewarded the dog. We recorded how long it took for the dog to figure out that one person was always generous. Then we switched the roles, so now the mean person was suddenly the generous person and vice versa. Again, we recorded how long it took for dogs to learn this new association.

We compared the results of dogs with those of chimpanzees and found that they were about the same when learning that one person was always generous. But when the generous person became mean, chimpanzees instantly noticed the switch, while dogs took much longer. It seems chimpanzees and wolves, rather than dogs, are remarkable when it comes to associative learning.

What we have seen in this chapter is that the same animal who is an extraordinary communicator is surprisingly dense when it comes to navigating through space or understanding the rules that govern the physical world. Of course, dogs might slowly learn to solve these kinds of problems through associative learning. But compared to wolves, rats, and chimpanzees, dogs are unimpressive when they are left to solve problems on their own. Dogs start to look extraordinary again when they are back where they belong – running with the pack.

8

PACK ANIMALS

Dogs work best in a social network

Dogs are not meant to be lone wolves. They struggle to solve a variety of cognitive problems on their own and do not seem to grasp simple laws of nature that even infants understand. Instead, the communicative abilities of dogs make them more suited to a social way of life.

One of the benefits of social living is the potential to learn new things from others that you would not discover on your own. Another benefit is the ability to co-operate and gain strength in numbers. Many animals learn from one another and co-operate, and the cognition involved can be simple or it can require a rich understanding of others. Perhaps there is more genius to discover in how dogs run with the pack.

Peer Pressure

In the summer, my brother brings his dog Cousin Carter to play with our dog Tassie at the lake. Carter and Tassie are about the same age, but since Carter is a Labrador retriever, he wanted to swim as soon as he could walk. Carter's favourite game is to chase tennis balls in the lake. As a puppy, Tassie had a more reserved approach to water. He tiptoed in; if you threw a ball, he might fetch it as long as he did not get his head wet.

But Tassie has a serious competitive streak. The first time we took Tassie and Carter down to the lake, seeing Carter belly-flop into the water and return triumphantly with the ball was too much for Tassie. Soon Tassie was belly-flopping into the water before I even threw the ball to get a head start on Carter.

Since then, Tassie has had no problem getting wet. Of course, maybe swimming was a skill he would have developed on his own, but the way his behaviour changed so immediately makes me think that it was Carter's influence.

Dogs can certainly be influenced by watching others. One of the basic skills animals must acquire is how to tell good food from bad. For instance, if a rat smells a new odour on the face of a recently deceased rat, they will avoid any food that has the same odour (making most rat poisons useless). Meanwhile, rats are attracted to odours of new foods from the mouths of healthy rats.

To test whether dogs learn about food in a similar way to rats, an experimenter gave dogs a choice between a new food that was flavoured with either basil or thyme. Initially, the dogs had no preference for either food. But if the dogs first encountered another dog who had eaten one of the two flavours, they tended to prefer the same flavour. If your dog is having problems eating new types of food, being in a social setting can help them transition to a new diet.

Not only do dogs decide what to eat from interacting with others, they also decide how much to eat. In another experiment, dogs were either fed alone or in front of other dogs who were eating. Dogs ate up to 86 percent more food when they could see other dogs eating than when eating alone. So if your dog needs to cut down on his calories, you might want to reserve a table for one.

By watching others, dogs might be able to spontaneously solve problems they struggle to solve on their own. We have seen wolves find their way to food around barriers faster and with fewer errors than dogs. But Peter Pongrácz from Eötvös Loránd University in Hungary thought that since dogs are pack animals, they might learn by watching other dogs avoid danger or obstacles, instead of working it out on their own. Pongrácz predicted that dogs would find their way round a detour much faster if they could observe someone else solving the problem first.

Dogs were given a series of problems requiring them to detour around a V-shaped fence. This fence was formed by two legs of three metres each, and food was placed inside the V.

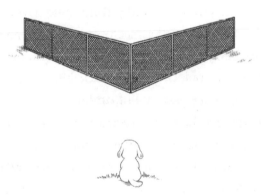

When alone, dogs struggled to round the edge of the fence. They would pace back and forth for as long as thirty seconds before realizing how to get the food. Even after repeatedly solving the detour, their speed did not improve.

Everything changed when the test became social. When the dogs saw either a dog or a human solve the problem first, they took a direct path around the obstacle on their very first trial, in less than ten seconds.

This means that dogs are sometimes faster at solving problems when they observe someone else's success rather than experiencing the same success themselves.

Is Your Dog a Copycat?

Lloyd Morgan's dog, Tony the terrier, discovered how to open a gate through trial and error. Tony became a classic example of how complex behaviour can be explained by simple forms of cognition. Tony had to use trial and error because he did not understand that he had to break the connection between the gate and the fence by moving the latch. But there was another way Tony could have learnt to open the gate immediately – if he had seen someone else open it first.

Even though dogs may be unremarkable in understanding connectivity while on a lead, they are remarkable at learning solutions by watching others. Dogs may not understand how it works, but they can see how you did it. In a recent study, food was hidden in a box behind a door. The door could be pushed to the left or right. Dogs were given the opportunity to open the door and get the food. Even though they were rewarded for pushing the door in either direction, from the very first trial, dogs moved the door in the same direction in which they saw it demonstrated by another dog.

In a similar study, however, dogs did not spontaneously copy a human. The dogs saw a human move a handle in a certain direction to release a toy. Even though the dogs were more attracted to the handle after the human touched it, the dogs did not move the handle in the same direction.

Though dogs may not spontaneously copy us, we can slowly condition one to do so. Experimenters conditioned a group of dogs to push open a door. Half of the dogs were rewarded for using the same method as a human, while the other half were rewarded for using a different method.

The dogs rewarded for copying the human learned much faster than the group of dogs who were rewarded for behaving differently. It seems that dogs have a natural bias towards copying someone else's actions. This may make it difficult for dogs to learn a task if you are trying to teach them to do as you say and not as you do.

Research suggests that dogs can spontaneously copy a single action of another dog, and in some cases they have a bias towards copying a human. But many problems require more complex solutions than detouring round a fence or moving a door to the left or right.

Even as infants, humans can copy a sequence of actions to solve a problem. To test whether dogs can also copy a sequence of actions, József Topál from Eötvös Loránd University recruited the service dog Phillip, whom we have already met.

Phillip was trained to perform three actions in response to three different verbal commands. Then, instead of giving Phillip the verbal command for each action, Topál would tell Phillip, 'Do as I do!' and act out an action. Even though Phillip only had Topál's behaviour to decide which of three actions to perform, Phillip rapidly learned to copy Topál's behaviour.

To see if Phillip could generalize his copying skills, Topál challenged Phillip with a combination of new actions. Phillip was often able to copy the new actions spontaneously, even though he sometimes had to compensate for the differences between himself and

Topál. For instance, when Topál spun on two legs, Philip spun on his four.

In a final amazing act, Phillip watched as Topál moved various objects back and forth from different positions. Even though Phillip had never been asked to solve this kind of problem before, when Topál said, 'Do as I do', Philip did not hesitate. Phillip moved the objects between the different positions the same way Topál had.

Dogs can spontaneously copy the solution to a problem – at least if it only requires a single action. And at least one dog, Phillip, can be trained to copy more complex sequences. This raises the question whether dogs make inferences when they copy the actions of others. For instance, say you watch someone assemble IKEA furniture and, in the middle of assembly, they scratch their nose. Based on their demonstration, when it is your turn to assemble IKEA furniture, you do not scratch your nose halfway through. You make an inference that nose scratching is irrelevant to building furniture.

In an experiment, infants watched an adult turn on a lamp with her head. The infants inferred that there was a reason why the adult did not use her hands to turn on the lamp. So they copied the adult's method of turning the lamp on with her head. However, when the adult again used her head but had a blanket wrapped around her arms and could not use her hands, the infants no longer used their head – they just switched on the lamp with their hands. The infants inferred that the adult did not use her hands because she could not use them. So the infants ignored the strange method and used their hands to turn on the lamp.

In a controversial finding, one study suggests dogs make the same inferences as infants. A dog was trained to demonstrate an awkward technique to other dogs – to pull a cord that released food with her paw.

When dogs saw the demonstrator do this, they faithfully copied the demonstrator, even though it would have been easier to use their mouths. But if the demonstrator had a ball in their mouth, which forced them to use their paw, dogs were more likely to use their mouths.

This result is remarkable because it is analogous to the finding with infants. It is also controversial because so far, no one has been able to replicate the results.

Dogs are clearly dependent on pack power. Although dogs are unlikely to solve a problem on their own, living in a group increases the chance that someone will discover the solution – even if by accident. Once someone stumbles on the solution, the rest of the pack can quickly learn from success.

This does not mean that dogs learn socially like humans. Provided that dogs are able to make social inferences as they decide what to copy and what not to copy, they are not as flexible or skilled at copying actions as we are. This is a blessing in many ways. If dogs could learn to use tools by watching us, imagine the locks we would have to put on our doors!

Co-operation in the Wild

Many dogs do not live as part of human families, and some have little or no direct contact with people. Feral dogs are dogs who have been domesticated but have returned to a wilder existence. These dogs include those who live completely independently from humans, as in the case of dingoes and New Guinea Singing Dogs, as well as stray dogs who survive by scavenging human refuse. Many populations of feral dogs have not been intentionally bred by humans for generations. By studying feral dogs, we can see how dogs make decisions without human intervention.

In some ways, feral dog packs are organized similarly to wolf packs. Feral dog packs usually have just a few dogs but can reach a stable size of more than ten. This is similar to wolves, whose typical pack size is usually fewer than seven but can reach up to thirty where food is abundant.

Researchers have observed dominance hierarchies in some feral dog packs. In these packs, adults are typically dominant to sub-adults and juveniles. Like wolves, dogs have favourite friends in the group and spend more time socializing with them.

The clearest signal that dogs and wolves value others in the group is their tendency to make up after any type of conflict. Both species usually rush to reconcile within minutes after a brawl or even a growl. Wolves are more likely to reconcile after a fight if they have a strong bond. Dogs are more likely to reconcile after a fight with a familiar dog rather than an unfamiliar dog. Having a positive inter-action shortly after a conflict is a powerful mechanism to re-estab-lish the social bonds that facilitate co-operation.

Feral dog packs differ from wolf packs in important ways, how-ever. While a feral dog pack is a mixture of unrelated dogs, members of a wolf pack are usually closely related. Just like people tend to prefer to help their family members rather than strangers, animals also tend to prefer to help family members. With fewer family mem-bers in their pack, feral dogs approach co-operation differently from wolves.

In wolves, with the exception of unusually large packs, a single breeding pair is dominant to everyone else. This pair uses their dominance to suppress the breeding of other pack members. Dom-inant female wolves are aggressive all year round and use unpro-voked attacks to prevent other females from mating. Even if a subordinate female manages to breed, the dominant female will often kill her pups. Male wolves become most aggressive during the mating season. They are especially likely to attack if they see another male attempting to mate with their favourite female. This may be something similar to jealousy.

Younger and subordinate pack members are usually the offspring of the breeding pair from previous years. Juveniles are forced to stay with their parents because meeting another wolf pack before they are fully grown would be dangerous. To earn their keep, juveniles help their parents raise the next generation. Juveniles bring back food to the pups after a successful hunt and protect the pups while their parents are hunting. Even though juveniles cannot have pups of their own, by contributing to the survival of their siblings, they indirectly pass on their genes. The breeding pair benefit from this co-operative arrangement since their elder offspring can continue to mature within the safety of the group while increasing the chance that their newest offspring will survive as well.

Feral dogs have a different system. While some feral dog groups have a dominance hierarchy that predicts priority of access to resources such as food and mating partners, this hierarchy is not as strict as in wolves. There is no dominant pair that leads the group. Instead, the leader of a feral dog pack is the dog who has the most friends. When the pack decides where to go, they do not follow the most dominant dog; instead, they follow the dog with the strongest social network.

In feral dogs, there is no co-operative breeding and no single breeding pair. Instead, dogs are promiscuous. Females mate with multiple males and rarely bond with a mate. Unlike wolves, the dominant females do not seem to prevent subordinate females from breeding.

Sexual freedom comes with heavy responsibility in the case of

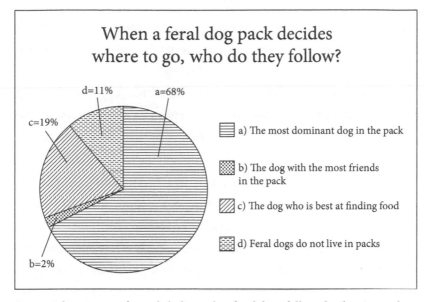

When a feral dog pack decides where to go, who do they follow?

d=11% a=68%

c=19%

a) The most dominant dog in the pack

b) The dog with the most friends in the pack

c) The dog who is best at finding food

d) Feral dogs do not live in packs

b=2%

Sixty-eight percent of people believe that feral dogs follow the dominant dog. Only 2 percent of people correctly answered that the dog with the most friends is the leader of the pack.

feral dogs. Without a bonded mate or suppressed juveniles, feral dog mothers have little help rearing their puppies. There are no helpers to provide food. This results in extraordinarily high mortality in feral dog puppies, and less than 5 percent survive their first birthday.

Even though feral dogs do not co-operate while raising their young, researchers were interested in whether they co-operate during conflict. Pack living is not always peaceful, and both wolves and dogs have their fair share of conflict. In one captive wolf pack, three brothers worked together to overthrow their father as the dominant male in the group. Although similar coalitions are difficult to observe in feral dogs, pet dog puppies will join one another in playful attacks on other puppies. If two puppies are having a play fight, when a third dog joins in they will almost always play-attack the dog who is losing. This teaming up is probably inherited from wolves since puppies jointly attack others while playing.

One of the main benefits of living in a pack is that you can protect yourself by co-operating to defend a territory. Wolves are one of the few animals who attack and kill members of their own species. If a group of wolves detects a lone wolf, they will try to catch and kill her. In an extreme case, in Denali, Alaska, an estimated 40 to 65 percent of wolves were killed by other wolves. No wonder juvenile wolves prefer the protection of a pack, despite not being able to breed.

Feral dogs are also territorial, although no one has ever reported them killing one another. Observations show the power of strength in numbers, even if the danger dogs pose to one another is minuscule compared with some wolf populations. When two feral dog packs meet, they tend to have a barking competition. Eventually, a member of one group leads the pack in a charge that typically scares off the other group. Physical fighting is rare when compared to wolves, but there is still a significant risk of injury. Researchers were curious about what made a pack go on the offensive and what determined the other pack's response. The researchers found that in most cases, group size predicted the winner. The larger pack was more likely to chase away the other pack. So for both wolves and dogs, staying together as a group is a crucial strategy for outcompeting other packs.

Another difference between the co-operative behaviour of wolves and dogs is how they find food. Wolves can flexibly adapt their hunting strategies depending on the available prey and hunting partners. Individual wolves or small groups can make a living on a variety of smaller prey species. Large groups of wolves are thought to co-ordinate with one another, in a similar way to lions, to bring down moose in Canada or musk ox in Mongolia.

In comparison, feral dogs are poor hunters. Feral dogs rely on human-made sources of food such as rubbish dumps. When feral dogs hunt, they are not usually successful. The exception is dogs who prey on species that did not evolve with large mammalian predators, like many marsupials of Australia. When researchers watch feral dogs hunting, there is nothing that looks like the co-ordination seen in other mammals.

This might be surprising, since dogs seem to be so useful to human hunters. Dogs have been bred and trained to work with people to detect and occasionally capture prey. Many modern breed families, such as retrievers, hunt with people using modern weapons, such as guns. These breeds tap into the genius of dogs by using human gestures and vocalizations to co-ordinate their activity.

Dogs who hunt with people using older technology show less co-ordination, however. In Nicaragua, the Mayangna and Miskito peoples are famous for hunting with dogs that resemble 'generic' village dogs in build and behaviour. The Mayangna and Miskito hunters simply allow the dogs to run freely without any human direction, training, or intentional breeding. Multiple dogs increase the likelihood that they will detect prey such as a tapir, but the dogs do not co-ordinate with one another. Instead, the Mayangna and Miskito hunters follow the barking that indicates that one or several of the dogs have found or cornered their prey. The hunters then shoot the prey with a dart or arrow. Modern sight hounds use a similar hunting strategy: They find and chase prey but do not co-ordinate with humans.

Dogs benefit from a pack lifestyle. They compensate for their

Co-operative behaviour of wolves and dogs compared

Form of Co-operation	Wolf	Feral & Pet Dogs
Protect territory	**Yes:** including lethal aggression against non–group members	**Yes:** no lethal aggression; rare physical contact with disputes typically decided by barking contest
Form coalitions	**Yes:** Mating pair work together to suppress others' reproduction; over-throw of male in mating pair	**Yes:** observed mainly during play
Co-operative puppy rearing	**Yes:** Juvenile offspring and male of mating pair provide food to pack puppies.	**Rare:** a few instances observed of male sharing food with puppies; no help from previous offspring
Hunting	**Yes:** Large groups can bring down even largest mammals by acting together: No experiments have tested the ability of wolves to co-ordinate their efforts or actively recruit help as done by other co-operative hunting species (e.g., hyenas and chimpanzees)	**Only with humans:** feral dogs primarily scavenge or focus on small prey (the exception may be the dingo); dogs only skilful at locating large prey during human-led hunts
Reconcile after fights	**Yes:** reconcile within 2 minutes of dispute and fastest with most important social partners	**Yes:** reconcile quickly and most often with familiar rather than unfamiliar dogs
Third-party consolation after fight	**Yes:** Those not involved in fight will affiliate with or 'console' those who fought.	**Yes:** Unlike some primates, a third party not involved in fight tends to console loser of fight more than winner.
Show preferences for partners	**Yes:** Wolves become jealous if pack members affiliate with their favoured social partner; unknown if they recognize different levels of skill, playfulness, or generosity in novel social partners.	**Yes:** Feral dogs follow preferred partners, not dominant dogs, when making travel decisions; pet dogs spontaneously prefer generous and playful human partners.
Co-operate with humans	**No:** relatively uninterested in humans even if raised by people	**Yes:** When raised by people they rely on humans to solve problems; can be trained to help people hunt, herd, and travel. More recently can be trained to help people with physical and mental disabilities or in detection of disease, bombs, trapped people, and more.

inability to solve problems on their own by watching how others solve the same problems. Dogs also defend the pack by sticking to-gether when they meet other groups. Wolves show more evidence of coalitionary behaviour and co-operative hunting than feral dogs. Given the particular nature of dog co-operation, researchers have conducted experiments to understand the cognitive abilities that allow for the forms of co-operation observed in dogs.

Can Dogs Spot a Cheater?

Feral dogs rely on co-operation to survive. They show a variety of co-operative behaviours as they interact with group mates and defend their territories. To understand what cognitive skills allow for dog co-operation, researchers have studied how dogs recognize and remember their social partners and how they evaluate new social partners. Researchers have also studied how dogs decide when to co-operate and who to co-operate with to avoid being cheated. Most controversially, researchers have begun asking whether dogs are motivated to co-operate out of concern for others or a sense of fairness.

To co-operate, you have to recognize potential co-operative partners and know a cheater when you see one. To do both, you have to be able to recognize individuals. Dogs seem to have an excellent memory of their friends. In Greek mythology, Odysseus's dog, Argos, famously recognized him after twenty years. Darwin wrote that his 'savage' dog, Czar, did not growl at him when he returned from his three-year voyage round the world on the HMS *Beagle*. Between my visits to Australia, Cina, whom I raised as a puppy, greeted me with the same excitement she saves for her owner even when I had not seen her for years.

Researchers have conducted several experiments to confirm whether dogs remember individuals. One experiment showed that dogs and their mothers recognize each other after being separated for two years. When dogs could choose to approach either their mother or a female of the same age and breed, they strongly preferred to approach their mother. Similarly, mothers were more likely to approach their offspring. Dogs also preferred a cloth that smelled like their mother to a cloth that smelled like another dog.

Surprisingly, dogs could not recognize their littermates after the two-year separation unless they had been living with them. This suggests that dogs can at least recognize their mothers and offspring when making decisions about whom to co-operate with in their social groups. They also can potentially recognize siblings if they continue living with them.

However, a pet dog's most important social partner will probably be a human. In terms of how dogs recognize people, they obviously use scent, but they can also integrate audio and visual information. Dogs were played a recording of the voice of their owner or a stranger. Then the dogs were shown a picture of their owner or a stranger. When the picture did not match the voice, dogs looked at the picture longer, as if they were surprised. The dogs had formed an expectation about the picture they would see based on the voice they had heard. They could only form such an expectation if – when they hear their owner's voice – they also remember what their owner looks like. The dogs could be doing this only by making an inference.

The next step in co-operation is to distinguish good co-operative partners from bad partners. Dogs watched people interacting with one another or with other dogs before choosing whom they wanted to interact with. In one condition, the first person shared food with someone, and the second person stole food from someone. In another condition, the first person allowed a dog to win a tug-of-war battle with a rope, and the second person did not let the dog win.

In both situations, dogs immediately preferred the generous human who gave food and the nice person who allowed the dog to win the tug-of-war. Dogs do not need to interact with potential co-operative partners to form an opinion about them. They can evaluate co-operative partners just by watching them play, compete, or even share food with others. Dogs seem skilled at detecting which individuals will be the best co-operative partners.

It is especially fortunate that dogs can detect and remember bad co-operative partners because feral dogs have been seen to 'cheat' during territorial disputes. These cheaters stay towards the back of the pack during disputes to minimize the risk of injuries. Because dogs can recognize different individuals and assess who is co-operative and who is not, they might avoid being repeatedly cheated.

To co-operate successfully, you have to recruit help when you need it. In one study, dogs were given a box of food that they could open, but then researchers fiddled with the box so the dogs could no longer open it. The dogs quickly looked to a human as if requesting help instead of continuing to try to solve the problem on their own. Dogs also use 'showing' behaviour to direct humans towards items they need help in obtaining. These studies suggest that dogs are good at recruiting help when necessary.

However, while dogs can recruit help, there is currently no evidence that they know who is the best person to recruit help from. If you need your oil changed, you do not take your car to the dry cleaner. When dogs were given a test where success depended on recruiting the right person for the job, dogs were not very good at doing so.

Another important skill for co-operation is the ability to know how many partners you need. Several experiments suggest that dogs can estimate quantities. In one, dogs were given a test that has been used to show that five-month-old infants have basic counting skills. Dogs watched as an experimenter hid either one or two dog treats behind a barrier. When the barrier was removed, sometimes there were more or fewer treats than the experimenter had hidden. Dogs stared longer when the number of treats was different from what they had seen hidden. They must have expected to see the same number, which means they were comparing the number of food pieces they were seeing with the number they remembered the experimenter hiding.

Another experiment let dogs choose between two plates that had different amounts of the same food. The greater the difference in the quantities, the easier it was for the dogs to choose the larger portion. For instance, dogs had no problem choosing between five and two pieces. It got more difficult when the quantities were similar, for instance when they had to choose between three and two pieces.

This research suggests that dogs can at least estimate quantities and may even be capable of basic counting. In an exciting twist, these studies directly map onto what Roberto Bonanni of the University of Parma observed in Italian feral dogs. Bonanni and his colleagues have observed how feral dog groups approach and flee during conflicts between different groups.

Since the larger pack usually wins a territorial dispute, dogs often join forces. If dogs know larger groups are safer, they should be bolder if their pack is bigger than another pack. Bonanni observed that it was almost always a member of the bigger pack who went on the offensive. This suggests that dogs understand there is safety in numbers and can evaluate the relative size of different groups.

Also, when Bonanni observed that both packs were small (fewer than four dogs) or when one pack was much larger than the other, members of the smaller pack never made the mistake of going on the offensive. Dogs were less cautious when both packs were larger and more similar in size. This suggests that while dogs do not do complex mathematics, they do keep track of when the odds are in their favour based on the relative size of the opposing pack. Dogs seem to know that when it comes to holding their ground, they have to work together to win.

All this research indicates that dogs have many of the basic cognitive skills to recognize and remember cheaters, to recruit help when necessary, and to know how many partners are needed.

Since dogs seem to have the cognitive skills that allow for co-operation, researchers wondered what motivates their co-operative behaviour. Perhaps dogs are motivated by guilt, empathy, and fairness in a similar way to us.

Guilt is a feeling of remorse for committing an offense or violating a social norm. More than 75 percent of owners believe their dog feels guilty for disobeying. The way dogs cower and slink away when you catch them doing wrong certainly gives a good impression of guilt. If dogs do feel guilt, it could motivate them to obey and co-operate with us.

Alexandra Horowitz of Barnard College conducted an experiment to see whether dogs can feel guilty. An owner told their dog not to eat a treat on the floor. Then the owner left the room. When the owner returned, an experimenter told them whether their dog had taken the treat or not. The trick was that in some trials, the experimenter told the owner the dog had eaten the treat when the dog had not. In other trials, the experimenter told the owner the dog had not eaten the treat, when the dog had.

Regardless of whether the dogs had reason to feel guilty or not, they always showed 'guilty' behaviour if the owners were told they had disobeyed. This is probably because owners used a disapproving voice or scolded their dogs. At least in this experiment, it does not look like the dogs were feeling guilty. They were just reacting to the owner's behaviour regardless of their previous actions.

Researchers have also debated whether animal co-operation is motivated by a sense of fairness, as it is in humans. Do animals have a sense of fairness that informs their emotional responses to co-operative outcomes? Researchers decided to test this by paying dogs for 'giving their paw'. Dogs were asked repeatedly to give their paw. Researchers measured how fast and how many times dogs would give their paw if they were not rewarded. Once this baseline level of paw giving was established, the researchers had two dogs sit next to each other and asked each dog in turn to give a paw. Then one of the dogs was given a better reward than the other. In response, the dogs who were being 'paid' less for the same work began giving their paw more reluctantly and stopped giving their paw sooner. This preliminary finding raises the intriguing possibility that dogs may have a basic sense of fairness, or at least an aversion to inequality.

Empathy can also motivate co-operative behaviour. If you see a child crying alone in a train station, you feel compelled to help. If you see a puppy whimpering after being bullied by another dog, again, you feel moved to help. As humans, we literally feel the pain of others. If we watch someone suffering, we experience negative emotions in response.

Researchers have looked for evidence of empathy in a range of animals. While it is difficult to know if an animal feels the pain of someone else, there is evidence that animals at least react to the pain of others. Mice who are near suffering mice show increased signs of suffering themselves. Bonobos and chimpanzees will often hug or kiss someone who was bullied.

One of the main arguments for empathy in animals is the display of consolation behaviour. After two chimpanzees fight, usually one chimpanzee will approach the other, and they will reconcile by grooming, hugging, or kissing. However, after some fights neither side is willing to reconcile. In these cases, a third chimpanzee who was not involved in the fight, but is typically a friend or family member, will approach and console one or both chimpanzees by grooming or hugging. Consolation behaviour is thought to be a powerful method of reducing tension in the group and preventing future fights.

Two studies suggest that both wolves and dogs display consolation behaviour. Consolation behaviour in dogs is particularly striking. Dogs prefer to console whoever loses the fight and submits, rather than the winner. In half of the consolations, the third party who was not involved in the fight actively initiated the contact with the victim by approaching. In many cases, this occurred even though the consoler did not witness the fight. It appeared that the dogs were reacting to the whimpering of the victim. Just as a human might comfort a puppy bullied by another dog, it seems dogs will do the same.

In order for dogs to feel empathy for people, they would need a way to recognize various human emotions. Dogs can recognize when we are upset from our voices, but researchers were curious whether they could recognize the more subtle visual cues of facial expressions.

A group of dogs was trained to always choose the photo of their owner smiling instead of the photo where their owner was not smiling. Then the researchers introduced twenty pairs of photos. Each pair had someone smiling and not smiling. The dogs were able to transfer their trained knowledge of their owner's face to a variety of unknown people. There was just one catch: the dogs could only distinguish a smiling face from a non-smiling face if the person was the same sex as their owner.

At the very least, it seems that dogs can rapidly learn to discriminate human facial expressions that are associated with differential outcomes – either positive or negative. Curiously, though, dogs may transfer what they have learned about their owner only to other people of the same gender. While this shows dogs can distinguish men from women, dogs might struggle to predict the behaviour of people of a different sex from their primary caretaker.

True empathy means you experience a negative feeling when someone else is suffering and a positive feeling when someone else is happy. It is almost as if we can catch other people's feelings contagiously. One measure of this social contagion is yawning. When someone yawns (or even as you read 'yawning') you are more likely to yawn in response. This is known as contagious yawning. Some have suggested that contagious yawning is related to our ability to respond to the emotions of others. Contagious yawning is related to empathy scores in adults, while many autistic children, who struggle to recognize the emotions of others, do not contagiously yawn.

Researchers looked at whether dogs contagiously yawn. More than 70 percent yawned in response to the yawning of an experimenter. Dogs yawned much less in a control condition where experimenters opened their mouths but did not yawn. This raises the possibility that dogs are sensitive to what others are feeling and may even feel similarly through contagion. Such extraordinary findings need to be replicated by other researchers so that we can better understand whether they really indicate that dogs can feel our pain.

Dogs are truly social creatures. Not only are dogs tolerant enough to live in a group, but they thrive by watching others solve problems they could not solve on their own. Dogs are also talented co-operators. From what we can see, dogs know when they need to co-operate, can recognize potential co-operative partners, and can tell a good partner from a bad one. Dogs are at their best when they can work together, both with us and with the rest of their pack.

PART THREE

YOUR
DOG

9

BEST IN BREED

The question on everyone's lips –
which breed is cleverest?

In 1994, word began to spread round dog parks that according to a scientific study, Border collies (followed by poodles, German shepherds, and golden retrievers) were the most intelligent breed. Owners of Border collies preened, and litters of Border collies began selling like hotcakes.

Usually, the first thing you know about your dog is the breed. Or if he is a mutt, you quickly learn to say he is part this and part that. Breeds are what get conversation going at the park. I would not mind if that changed.

What fascinates me about dogs is that all have their own unique intelligence. All dogs can follow social cues, but some are extraordinarily good at it. Other dogs are better at making inferences, or understanding gestures, or navigating. I am not so interested in fancy tricks and what dogs can be trained to do. Instead, I love seeing what dogs do when they see a problem for the first time. What do they understand about the problem? What kind of intelligence will they display? What skills will they use to solve it?

191

At the park, I think it would be so much more interesting if, instead of swapping information about breeds, people would share the unique intelligence of their dogs. Because that is what makes their dogs really special.

There are various methodological problems that make it difficult to find breed differences in cognition. The first is that no one can agree what a breed is. The American Kennel Club recognizes 170 breeds, the Kennel Club of the UK recognizes 210 breeds, while the Australian National Kennel Council lists 201 – and that is just in the English-speaking countries. In total, more than four hundred breeds are recognized by kennel clubs all over the world.

The main criteria used for most breeds today is appearance. A retriever who does not retrieve but looks like a retriever is still classified a retriever. A sheepdog who does not herd sheep is still a sheepdog. Because breeding is not based on specific behaviours, it is difficult to predict which breed will have more or fewer cognitive skills.

This was not always the case. Breeds as we know them today are a modern invention – most have been around for only a few hundred years. Historically, dogs were divided into breeds based not on their appearance but by function, so any dog who chased hares was a harrier, any lapdog was a spaniel, and any large, intimidating dog was a mastiff.

Over time, the emphasis on a breed's function began to favour a certain appearance. For instance, in a barbaric practice, eighteenth-century English butchers were required to tie a bull to a stake and unleash a group of dogs to kill it, as this supposedly made the meat more tender. Bullbaiting, as this was called, quickly became a popular sport with gamblers.

Any dog with a temperament for launching itself at an enraged bull was called a bulldog, but in general, it was helpful for the dogs to be low to the ground, since bulls charged with their heads down. The dogs had to have strong jaws so they could lock onto the tender nose of the bull and not have their teeth torn out when the bull flung them around. To breathe during all of this, it was advantageous to have a protruding mandible and wide, flared nostrils. Selection for these traits likely shaped the bulldog into the breed we know today.

In the nineteenth century, it was the aspiring new professional class who turned dog breeding into England's national obsession. Insecure about lineage and social standing, they did not want to have just any mutt on the end of their lead. They wanted people to know at a glance that they had a first-rate dog who had cost a lot of money and had impeccable lineage. The easiest way to broadcast this was by the dog's appearance.

Initially, the sporting gentry were appalled. The focus on appearance, rather than on function, could ruin the dogs they used for hunting and other sports. Only skilled hunters could train and handle a sporting hound, but pet dogs could be owned by anyone.

Perhaps to appease the opposition, the first formal dog show was targeted at the upper-class sporting gentry and was held on 28 June 1859 – the same year that Darwin published *On the Origin of Species*. In this landmark event, there were only two classes: pointers and setters. Four years later, the show exploded, with more than a thousand entries. Dog shows became the place where the nouveau riche came to throw their money around. While you could buy a sheepdog for a pound, a first-class show collie was worth up to a thousand pounds, which translates roughly to £75,000 today.

As you can imagine, this kind of money attracted some unscrupulous people. With a little trimming and shoe polish, a sheepdog might be passed off as a show collie, and by the time such a scam was discovered, the perpetrators were long gone. In response to these swindles, the first kennel club was set up in 1873 to establish the identity and descent of pedigree dogs.

The Genetics of Dog Breeds

This means most of the breeds we recognize today originated less than 150 years ago. On an evolutionary time scale, this is a fraction of a nanosecond. Dogs and wolves are estimated to have split from each other between 15,000 and 40,000 years ago. Since then, only about 0.04 percent of the dog genome has evolved. This is a small

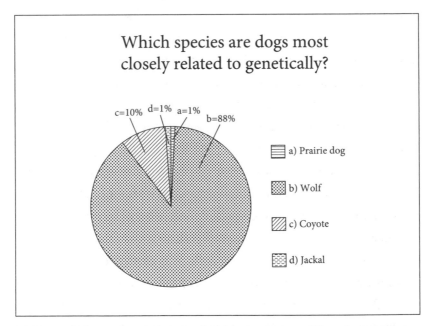

Most people know that wolves are the closest relatives of dogs, but one in ten people believes dogs are more closely related to coyotes.

difference when you consider that a dog is genetically 99.96 percent wolf.

With the publication of the dog genome in 2003, geneticists were finally able to confirm that dogs descended from wolves.

Geneticists could also classify the genetic relationships of all of the modern breeds of dogs. Based on these comparisons, we learned there are only two major groups of breeds.

The first group consists of nine breeds that are more genetically similar to wolves. These wolf-like breeds share more genes in common with wolves than other breeds. The striking thing about these more wolf-like breeds is that they are not from one geographic region. This is tentative support for the idea that dogs evolved multiple times in multiple places due to self-domestication. Besides the Middle Eastern group (Afghan hound and saluki), the dog most closely related to wolves, the basenji, comes from Africa. There are five Asian breeds: akita, chow chow, dingo, New Guinea Singing Dog, and shar pei. The two breeds from the Arctic – the Siberian husky and Alaskan malamute – are the only two breeds that show evidence of recent interbreeding with wolves. This makes sense, given they are the only breeds that live within the range of modern wolves.

The second group consists of the majority of modern breeds, which are lumped together into one group known as dogs of 'European origin'. While these dogs look and act differently from one another, very few genetic changes are responsible for these differences. After only 150 years of genetic separation, the differences among breeds of European origin are so small they are barely detectable.

Physically, dogs vary more than any other species on Earth, so you might think that a Chihuahua and a St Bernard would have different genetic profiles. But it appears only a small number of genes are responsible for the extraordinary physical variation in dogs. For example, think of all the different coat colours and textures of dogs, from the dreadlocks of the komondor to the tight curls of a poodle. Although we do not know exactly how many genes there are in the dog genome, there are probably tens of thousands – and the majority of coat types in dogs are regulated by just three of them. Breeders who focus on a dog's appearance can cause a major change in a breed but only affect a small number of genes responsible for this morphological trait. This is why breeds look so different but genetically are still so similar.

Given how genetically similar most breeds are and the fact that most breeds have largely been selected based on their appearance, our default hypothesis is that there will be few if any cognitive differences among various breeds. This might seem counterintuitive, given that certain breeds are famous for certain types of jobs, but scientifically, any cognitive changes caused by evolution must result from genetic change.

With European breeds being so closely related, geneticists keep changing their minds about the relationship among them. For instance, in 2004, German shepherds were grouped with the large mastiff-type dogs like the Newfoundlands and Rottweilers; in 2007, they were put in a new cluster – the mountain group, along with St Bernards – but by 2010, they were grouped with the working dogs, along with Dobermann pinschers and Portuguese water dogs.

Puzzle Pieces

Despite the constantly changing genetic landscape, you might have heard about DNA tests that can tell you what breed your dog is. Morgan Henderson is a graduate student in genetics at Duke University, and in 2010 she adopted a mutt called Roxy from a rescue home. She often wondered what kind of breed Roxy might be. Based on her appearance, her colour, and the way she used her tail to get attention, Morgan suspected she was a Nova Scotia duck tolling retriever. Reportedly these dogs are known for wagging their tails to attract ducks' attention and lure them close to hunters. So Morgan decided to get Roxy's DNA tested, but, being a geneticist, she put a little more thought into it than most people might.

She found that there are two types of tests: a cheek swab, which looks at the DNA in saliva and costs around £40 to £50, and a blood draw, which cost around £100 and has to be done by a vet. Morgan eventually decided on a blood draw, since, as a geneticist, she knew blood draws are usually more accurate.

DNA testing companies look at specific areas on the genome called markers, where there are predictable DNA differences between breeds. So at one marker on the genome, a boxer might have a DNA code like this: GGT, while at the same place in the genome of a Parson Russell, the code looks like this: GGC. So if you send away your dog's DNA to be tested and on that specific marker of the genome it says GGT, there is a high probability your dog has boxer in its ancestry.

One DNA company collects information on 321 of these DNA markers from each dog it tests. It has data for what DNA code each of 225 breeds typically has at each of these markers. This allows the company to calculate how closely your dog's code at each marker fits to the code of different breeds at these same marker sites. A computer then calculates how likely it is that your dog's DNA matches one breed or the other. These tests are by no means 100 percent accurate, and each dog can be a mix of many different breeds. The DNA test picks out the three breeds your dog most closely matches. Doggy DNA testing is being used for everything from determining a dog's heritage to profiling dogs in apartment blocks whose owners do not pick up their poop. Roxy's three dominant breeds were shih tzu, boxer, and basenji – something Morgan never would have guessed.

Genes, Behaviour, and Breeds

Long before it was possible to examine the entire genome of the dog, scientists were fascinated by the genetics of dog behaviour. In the early twentieth century, when excitement was increasing over the implications of Mendel and his pea plants, geneticists started look-ing for the same patterns of inheritance in dogs. The ultimate goal has been to understand the genetics driving behavioural differences we see in each breed.

Unfortunately, so far the challenge has been too great. First, most behavioural traits are not Mendellian traits controlled by one gene. Instead, behavioural traits are under the control of families of genes. Finding the genes in these families and the role of these genes is difficult even in fruit flies. In a complex, slow-breeding animal like a dog, it is nearly impossible.

Second, to understand breed differences in behaviour you would need to compare at least thirty dogs from each breed. They would have to be puppies raised and tested in a similar manner to control for the effect of rearing history and age on performance. If you took the Kennel Club breeds or all breeds worldwide, you would need between 6,000 and 12,000 puppies, decades of work, millions of dollars, and about a thousand graduate students. It is no wonder no one has done it.

The only person who has come close is John Paul Scott, who conducted the most comprehensive breed experiment of the twentieth century. At the end of World War II, when the American economy was thriving, the US government was pouring money into the sciences to compete with the USSR. It has often been called the Golden Age of Science. From 1945 to 1965, gifted young people were flocking to the sciences, and discoveries were being made every year.

Scott started his genetic work with fruit flies, but he was more interested in behavioural genetics – behaviour that could be passed down from generation to generation – and for that he would need to study mammals. Scott inherited a large, well-funded dog laboratory by chance (the project leader died), and in 1947 he hired John Fuller, a talented behavioural geneticist who had previously worked on mice.

Together, they undertook a brave new study, looking at the effects of genetics on behaviour. In the aftermath of World War II, people were still reeling from the Nazis' exploitation of eugenics and their idea of breeding a 'super-race', and so studies on the heritability of desired traits were not very fashionable. But, convinced of the application this could have to humans, Scott and Fuller persevered.

They did not have four hundred breeds and 12,000 puppies, but they had a good number – five breeds and 470 puppies. The breeds were chosen because they were about the same size, with 'normal' physique (no short legs) and wide behavioural differences. They bred a total of 269 dogs of the following breeds: basenjis (51), beagles (70),

American cocker spaniels (70), Shetland sheepdogs (34), and wire fox terriers (44). They also bred 201 hybrids to test the properties of Mendellian inheritance.

The purebred puppies were brought up in a strictly controlled environment. They were fed the same food, housed in identical runs, and began their training at the same age with the same tests.

Scott and Fuller ran a range of experiments from behaviour to intelligence. The results were published in what became a five-hundred-page bible for breeders, vets, and scientists called *Genetics and the Social Behavior of the Dog*.

The book is sort of a magic mirror in that it reflects the desires and beliefs of the reader. If you want to say Scott and Fuller found many significant breed differences, you can find evidence for this. If you want to say they found few breed differences, there is evidence for this, too.

The results were complicated. For example, to test emotional reactivity, Scott and Fuller restrained the dogs on a Pavlov stand, an apparatus where the dogs could be restrained and subjected to various situations that would elicit a response, such as electric shocks, loud noises, or an experimenter who 'grasped the dog's muzzle while speaking in a loud, harsh voice and forced the subject's head from side to side'. It is hard to say which breed was the most reactive. Basenjis were most likely to bite (83 percent), whereas most beagles barked (89 percent), and terriers were the most resistant to forced movement (53 percent). So it was unclear which measurement – biting, barking, or struggling – constitutes the most 'reactive'.

In terms of trainability, again, it was complicated. For instance, in a 'lead' test, where the dogs had to walk calmly on a lead, basenjis were terrible and resisted restraint but did not bark, whereas beagles were good on the lead but howled and wailed. Cocker spaniels balked at doors and gates, while shelties tended to leap on the handler and wind between their legs, getting into a complete tangle.

Scott and Fuller also used a series of 'intelligence' tests, but these were the same as those that the behaviourists had designed for pigeons and rats – things like mazes and detour tests. These tests gave no credit to the flexibility of dog cognition. Even so, no breed came out the clear winner or loser. One breed that did well on speed might do badly on accuracy.

From my reading of the magic mirror, Scott and Fuller basically found no breed differences, and they admitted as much when they wrote:

> After emphasizing differences between the breeds . . . we wish to caution the reader against accepting the idea of a breed stereotype.

The book is not exactly light reading, and it is now half a century old. It is an indication of the times that they could get away with referring to females as the 'weaker sex'. They also felt compelled to make the point that crossbreeding the dogs produced 'physically and behaviorally excellent animals', because at the time, the governor of the state of Alabama had commissioned a paper using reported 'physical and behavioral disharmonies' in hybrid dogs to caution against the dangers of crossbreeding races in humans.

Since Scott and Fuller, no one has had the money or the inclination to repeat the experiment on a larger scale. Apart from its being incredibly expensive to raise, house, and train almost five hundred dogs over ten years, there are the ethical considerations of what to do with five hundred dogs after the experiment is over.

Breed Personality

Just because it is difficult to find breed differences does not mean they do not exist. There are physical disabilities that are unique to some breeds, such as respiratory problems in pugs or a certain type of cancer in German shepherds. There are also psychological issues. For instance, some bull terriers have obsessive-compulsive disorder, where they chase their tails for hours a day.

Behaviour can also be inherited. In herding dogs, the Australian cattle dog herds by nipping at the heels of livestock, pushing them forward. Other breeds, like Border collies, get in front of the livestock, using the 'eye' to stare them down.

The question is, what kind of behaviours are conserved in breeds, and does each breed have its own 'personality'? Originally, dogs were bred for function rather than appearance, so it would be interesting to find out if these functional traits still exist.

Even in humans, personality studies did not begin to gain widespread acceptance until the 1980s. To measure personality, a team of psychologists collected all the words that were used to 'distinguish the behavior of one human being from that of another'. This amounted to some 18,000 words. They finally whittled them down into five major groups called the Big Five:

- Openness (artistic, curious, imaginative, with a wide range of interests)
- Conscientiousness (efficient, organized, responsible, ambitious, able to delay gratification)

- Extroversion (assertive, energetic, enthusiastic, finds the company of others stimulating)
- Agreeableness (forgiving, generous, kind, considerate towards others)
- Neuroticism (anxious, tense, sensitive to criticism, moody)

People were asked to assign scores to statements such as 'I have a vivid imagination' or 'I like order'; then the scores were tallied and a percentage was assigned to each of the Big Five.

Personality tests are useful because to a certain extent, you can predict the outcome of someone's life based on their personality traits. For instance, a high level of conscientiousness and openness usually means someone will do well academically. Low agreeableness and low conscientiousness in a child can predict antisocial behaviour. People with high extroversion tend to be successful in sales and management positions. Researchers have even found a link between personality traits and health, where people with high conscientiousness tend to live long, healthy lives, while high neuroticism tends to be a health risk.

Of course, personality is not the only factor in predicting what kind of life you have. It has to be combined with the right circumstances and the right (or wrong) environment. But personality tests can help identify those at risk, and certain traits can be manipulated in order to improve someone's health, happiness, and success.

When human personality tests gained prominence, researchers wanted to know if personality tests would work with animals, too. Dog personality tests have become popular in the last decade because predicting a dog's success at certain tasks would be helpful to us. Imagine being able to test which dog would make a good guide dog or companion for children and which, under certain circumstances, might be dangerously aggressive.

Because animals cannot talk, the application of a human personality test is a bit tricky. Human personality tests depend on vocabulary – not just the vocabulary that other people use to describe you, but the vocabulary you use to describe yourself. There is really no way to know whether your dachshund is imaginative with an appreciation for beauty or whether your poodle prefers to be organized and efficient. Nevertheless, researchers Kenth Svartberg from Stockholm University and Björn Forkman from the Royal Veterinary and Agricultural University in Denmark analysed a massive data set that included 15,329 dogs from 164 different breeds in an attempt to better understand dog personality.

The dogs were put in a series of situations, then specially trained judges recorded their reactions. Imagine how your dog would respond in these tests.

You and your dog are standing on a path in the woods. A stranger approaches. They shake your hand then pat your dog. The stranger then takes your dog for a short walk, about ten metres away. If all goes well, the stranger pulls out a rag and tries to play a little tug-of-war and fetch.

A little farther down the path, a small, furry object is pulled around a zigzagging course by a string. If your dog chases it, the mysterious furry creature flees, as if alive.

Then things start to get a little weird. More than thirty metres away, a stranger in a cape suddenly appears. His face is hidden by a hood as he creeps slowly towards you. He flaps his cape, like a bat, all the time creeping forward. If your dog does not attack the stranger, he pushes back his hood, revealing his face, and then tries to play with your dog.

You and your dog walk on. A dummy dressed in a boilersuit (the kind serial killers wear in Hollywood films) lies on the forest floor. Its feet are fastened to the ground; its arms are attached to two trees with ropes and metal loops in such a way that, when a hidden person pulls the ropes, the dummy suddenly stands upright. When you and your dog approach within three metres, the hidden person pulls on the rope, and the dummy leaps to a standing position.

Farther along the path, just as you pass, someone drags some chains across a sheet of corrugated metal, making a horrific racket.

Then comes my personal favourite. Two people dressed up like ghosts, white sheets and all, hide in the forest. They have white plastic buckets on their heads with holes for the eyes. Eyes and mouth are drawn on the bucket in black. When you are about twenty metres away, the ghosts slowly emerge from behind the trees and move in a ghostly fashion towards you and your dog.

To top it all off, at the end of the obstacle course, someone fires a gun.

You have to love the Swedes – it sure beats a boring questionnaire (although I hope they screened for heart conditions!). Throughout the tests, the dogs were rated on their reactions to the different situations: whether they were comfortable playing and interacting with a stranger, whether they chased the little furry object, whether they tried to kill the dummy, and so forth.

After putting more than 15,000 dogs through the tests, Svartberg and Forkman came up with five personality traits in dogs to correspond to the Big Five in humans. They were:

- Playfulness
- Curiosity/fearlessness
- Chase proneness

- Sociability
- Aggressiveness

First, they found that when it came to breed differences, there was so much variation within each breed that there were no real differences among the breeds. Second, in humans, the Big Five are not necessarily related. So you can have high conscientiousness but low extroversion. Or low agreeableness but high openness. But in dogs, the first four personality traits are related. So if your dog is playful, she is probably also curious, social, and likes to chase things. Svartberg and Forkman called the combination of these traits *boldness*. The other end of the spectrum was shyness. So shy dogs are timid, cautious, and evasive in social and non-social novel situations, whereas bolder dogs are more exploratory.

The only trait that was independent from all of the others was aggression. So your dog could be playful, curious, and social, but also aggressive under certain circumstances.

What was interesting about the shy–bold continuum is that just as the bold foxes who approached the human hand were more successful at using human social cues, when Svartberg tested Belgian tervuerens and German shepherds, bolder dogs did better on tasks such as obedience, searching, and tracking. They also were successful at these trials at a younger age.

The shyness–boldness trait has also been shown to be relevant in children: Those who are shy at two years of age and show avoidance of novel situations tend to be quiet and socially avoidant at seven years of age, while those who are bolder in novel situations are more talkative and interactive when older, suggesting that the shyness–boldness continuum is one that is conserved over time.

The shyness–boldness continuum is present in many other animals, including wolf cubs (wolf cubs who are bolder are more successful in killing prey).

Borbála Turcsán, from Eötvös Loránd University in Hungary, and colleagues used a tamer method to assign personality traits to different breeds. They had more than 14,000 dog owners fill in a questionnaire for their dogs that was adapted from a human personality test.

In addition to boldness, the researchers added three more traits: sociability (does your dog get on well with other dogs?), calmness (is your dog coolheaded even in stressful situations?), and trainability (does your dog learn quickly?). They also used the latest genetic data to assign each dog to one of five groups: the ancient breeds (breeds with ancient Asian or African origin, more closely related to wolves), and then the modern breeds: mastiff/terrier (mastiff-type breeds or breeds with mastiff-type ancestors and terriers), herding/sight hounds (breeds used as herding dogs and sight hounds), hunting (breeds with recent European origin, primarily different hunting dogs: spaniels, terriers, and hounds), and mountain (large mountain dogs and a subset of spaniels).

They found that among all of the breeds, the ancient breeds, which include chow chows, huskies, and basenjis, were the least trainable and the shyest, but also the calmest, which makes sense in light of their more wolf-like genetics. The mastiff/terrier cluster, which includes bulldogs, pit bulls, and mastiffs, were the boldest, while the herding/sight hounds such as Border collies and greyhounds were the most social and the most trainable. The least social group was the mountain group, which includes breeds such as cocker spaniels and St Bernards.

Personality factors compared across breed groups

	Most	Least
Boldness	mastiff / terrier	ancient breeds
Sociability	herding dogs / sight hounds	ancient breeds
Calmness	ancient breeds	mountain / hunting / herding dogs / sight hounds
Trainability	herding dogs / sight hounds	ancient breeds

Some of the findings using this approach contradict what several kennel clubs might have predicted about certain breeds. For example, according to the international kennel club Fédération Cynologique Internationale (FCI), the Spanish greyhound is not supposed to be overly shy, while Turcsán found that it was the shyest of all the breeds. Another example is Anatolian shepherds, which are penalized for aggressiveness in dog shows; but Turcsán found that they tended to be aggressive towards other dogs.

By thinking of breeds in terms of their genetic relatedness, and not just the way they have been historically organized by kennel clubs, Turcsán and his colleagues have made real progress in defining the characteristics of breeds, at least at a group level.

The Aggressive Breed Myth

Most of the research on breed differences has concentrated on one trait – aggression. The best estimates suggest that around 250,000 people in the UK and 4.7 million people in the US are bitten by dogs every year. The US Centers for Disease Control and Prevention (CDC) estimates that 885,000 of people bitten by dogs require medical attention, and more than 30,000 Americans required reconstructive surgery for dog bites in 2006. Most of these bites affect children, and one study found that half the children in the US have been bitten by the age of twelve, and more than half the children bitten by dogs suffer from post-traumatic stress disorder.

Children bitten by dogs cost US insurance companies $345 million a year; the total loss associated with dog bites may be as much as a billion dollars annually. Attacks in England are estimated to cost the NHS £3.3 million per year. Because of the sheer number of people affected, dog bites have been referred to as an epidemic. It is no wonder that if there is a breed of dog more dangerous than others, legislators, politicians, insurance companies, and parents want to know about it.

If you believe news reports over the last decade, you would likely think that pit bulls are responsible for most dog-related injuries and deaths. Pit bulls are not a specific breed, but the general name given to three breeds – the American Staffordshire terrier, the Staffordshire bull terrier, and the pit bull terrier (the American Kennel Club does not recognize the American pit bull terrier as a distinct breed).

Genetic work shows that pit bulls are genetically similar to bulldogs, so they probably shared a common ancestor who was once used in bullbaiting. Their reputation for fearlessness, courage, and unwillingness to back down from a fight makes these dogs desirable in activities such as dogfighting. Pit bull terriers who are not aggressive enough are killed in the ring or destroyed. In this way, lines of dogs can undergo selection for enhanced aggression – similar to how the Soviet scientists created a more aggressive line of foxes by breeding the foxes most likely to attack a human. These more aggressive dogs can end up mixing in with the general population and with the public.

After several much-publicized attacks in the 1980s, pit bulls became popular with people looking for a tough dog who could be trained to attack. Pit bulls are rumoured to have locking jaws and a bite force of 1,800 pounds per square inch (124 bar), although there is no scientific evidence to support either of these claims. In fact, Dr Brady Barr from National Geographic's *Dangerous Encounters* television programme tested the bite force of pit bulls, and found that their bite force (235 psi, or 16 bar) was considerably less than a Rottweiler's (328 psi, or 22.5 bar) and much less than a wolf's (400 psi, or 27.5 bar). There is no such thing as a locking jaw mechanism in pit bulls or any other dog.

Still, we devour newspaper stories about pit bull attacks. In the ten-year period between 2001 and 2010, there were 3,340 news articles on pit bull attacks, more than twice as many as on the alleged next most dangerous breed, German shepherds. According to an organization called DogsBite.org, through scouring the Internet for media reports, they found a total of eighty-eight US dog-related fatalities in the three-year period from 2006 through 2008. Pit bulls were responsible for 59 percent of these deaths. DogsBite.org also claims 94 percent of pit bull attacks on children are unprovoked, and three pit bulls are shot and killed every day because of aggression.

Various scientific articles seem to support the claim – one study found that 42 percent of dog-bite-related deaths reported between 1979 and 1988 involved pit bull types. Another found that pit bulls and Rottweilers were responsible for 60 percent of deaths from 1979 to 1996.

Occasionally, there is such a horrific attack that the public outcry leads to a push for strong legislation. This has resulted in breed-specific legislation mandating that pit bulls must be registered, spayed, neutered, muzzled, or banned altogether. For instance, the Dangerous Dogs Act of 1991 specifically bans owning, breeding, selling, or exchanging pit bull terriers (as well as the Japanese tosa, dogo Argentina, and fila Brasileiro), though the law is clever to call these 'types' rather than 'breeds' of dogs. The state of Ohio has also passed legislation regarding pit bulls, declaring that:

> The ownership, keeping, or harboring of such a breed of dog [pit bull types] shall be prima-facie evidence of the ownership, keeping, or harboring of a vicious dog. . . . A vicious dog must be confined on the owner's property by means of a locked fenced yard, a locked dog pen that has a top, or some other locked enclosure that has a top (such as a house). The owner must maintain at least $100,000 of liability insurance coverage on the animal.

These all seem like logical steps to protect children as well as the greater public from predictably dangerous dogs. The problem is that it is unclear if the pit bulls involved in the attacks are actually pit bulls.

In a 2009 study, researchers looked at how seventeen different adoption agencies classified breeds. A dog turning up at a rescue home is a much less stressful situation than a child turning up at an A&E, and you could argue that people who work with dogs are more experienced at identifying breeds than the average person.

Rehoming agency workers were asked to identify the dominant breed in several dogs. Blood samples from the dogs were then sent off for DNA analysis. Two-thirds of the time, the rehoming agency said the dog was predominantly a breed that was nowhere in the dog's ancestry. When the dominant breed was a Dalmatian, they called it a terrier. When the dog was mostly Alaskan malamute, they called it an Australian shepherd dog. If even experienced people who work with dogs full-time only get breeds right a third of the time, it is probable that the rest of us would get the breed wrong with an even higher frequency.

When hospital staff record that the dog who bit a child was a pit bull, they rely on the report of the victim, parents, or a witness. No one does a DNA test to make sure. Any dog with short hair, medium build, and a broad face might be called a pit bull. And Scott and Fuller found that sometimes dogs end up looking nothing like either parent. When they crossed basenjis and cocker spaniels, most of the puppies were unidentifiable except in that they were about medium build with splotchy coats. This means that looks can be deceiving. A dog that looks nothing like a pit bull may have pit bull genes, while a dog that looks like a pit bull is nothing of the sort.

In another study, people are more likely to perceive a dog as dangerous depending on the accessories being worn by the dog and its owner. A white male in his thirties dressed in a sports jacket, collar, and tie was photographed with a black Labrador with a matching lead and collar. Then another photograph was taken when the man was wearing torn, dirty jeans, an old T-shirt, and scuffed work boots while the dog was in a studded leather collar with a frayed rope for a lead. People judged the second photograph as depicting a more aggressive dog – even though they were pictures of the same dog.

In a review of other scientific studies, pit bulls are not even the main culprits. One analysis of eighty-four dog bites in children found pit bulls were responsible for 'a notable proportion' of bites, but this added up to only 13 percent. Another found that the top biting culprit was German shepherds, while English springer spaniels also ranked high. In a review of studies from 1971 to 1989, there is a sample of dogs that were found to be among the 'top three' biters: chow chows, collies, German shepherds, mixed breeds, American Staffordshire terriers, cocker spaniels, St Bernards, Lhasa apsos,

Dobermann pinschers, Rottweilers, poodles, and Nova Scotia duck tolling retrievers. The data is all over the place.

Then there is a question of the type of aggression – whether dogs are aggressive towards their owner, strangers, or other dogs. One study found that the most aggressive dog towards strangers and other dogs was the dachshund.

Because the research is inconclusive, instead of breed-specific legislation, seventeen US states have 'one bite' laws that hold owners responsible only after their dog has bitten someone. This is not perfect, as most dogs are first-time offenders, but it is better than imposing a ban on a breed of dog that probably is not even responsible for most of the injuries, creating a false sense of security among the public. We need to know more about behavioural aggression before we can hope to create laws that will have any effect on the dog-bite epidemic.

The rate of deaths from these bites is very low, and only 1 in 3.9 million dogs ever kills anyone. You are 573 times more likely to get killed by a car and 3 times as likely to get struck by lightning as to be killed by a dog.

While we are not sure about breed differences when it comes to aggression, what science *can* tell us is that 70 percent of bites happen to children under the age of ten. More than 60 percent of the children bitten are boys, and 87 percent are white. Children are most frequently bitten (61 percent of the time) when they come in contact with the dog's food or possessions. The children will usually be injured in the head and neck area, 55 percent to the cheek and lips, with the average length of the wound being 7.5 centimetres. Most of the breeds involved are large dogs, and male dogs are more likely to bite than female dogs. Two-thirds of the dogs who bite children have never before bitten a child, and between 25 and 33 percent of the dogs who bit were family pets.

In conclusion, the breed most likely to be involved in a serious dog-related injury is a child, usually a white male under the age of ten, with a large male dog living in the family home.

The Cleverest Breed

The question on everybody's mind is which breed is the cleverest. When it comes to cognition, there is almost no research on breed differences. This is surprising, since there is so much popular literature on what the most intelligent breed is. The lack of research comes mostly out of the fact that there are so many breeds and so little agreement on what a breed is. Only recently have we had data to help us group breeds based on their genetic relationship to one another.

Although at the moment, scientists have little to say about which breed is the most cognitively sophisticated, that does not mean people do not have strong opinions on the topic. When it comes to opinions, the top-ranking dogs are usually the same, although their order might change. Most people think that Border collies, German shepherds, retrievers, and poodles are among the cleverest dogs.

But the few data we have for evaluating breed differences do not support these choices. There were no breed differences in learning how to take a detour around a barrier based on a human demonstration. Other studies have found few systematic breed differences in the ability of dogs to follow a human pointing gesture. As comparisons of New Guinea Singing Dogs, dingoes, and rescue dogs have shown, all dogs are quite skilled at reading human gestures.

Only a handful of studies have reported breed differences in cognitive tasks. One of these studies is my research in Part Two comparing the ability of working and non-working breeds. While both groups of dogs performed well at reading human gestures, the working dogs were better at it.

As we have seen, dog genius depends much on communication. Monitoring a human's gaze is no doubt a valuable tool in a dog's cognitive tool kit. Mariana Bentosela of Buenos Aires University in Argentina and her team looked at the rate at which different breeds looked at their owner's faces. They found that retrievers gaze at experimenters' faces longer than German shepherds when trying to obtain a food reward. They suggest this may be because retrievers need to interact with their human partner to find where prey has fallen for them to retrieve. Shepherds guarding livestock may be less reliant on responding to humans, as their job requires more independence.

The problem with these studies, though, is that the ideal comparisons would be between groups of dogs who have been reared and tested in a similar way. Currently, the existing breed comparisons cannot rule out differences in rearing experiences that may explain the results.

Moreover, William Helton from the University of Canterbury in New Zealand has proposed another reason for finding breed differences that is unrelated to cognitive ability. Helton thinks the few breed differences people have observed have more to do with size of the breed than their history as working or non-working dogs.

Dogs can be dolichocephalic (having long skulls, like greyhounds), mesocephalic (having average skulls, such as Border collies), or brachycephalic (having broad skulls, such as Staffordshire bull terriers). The way a dog's skull is shaped has a lot to do with how the dog sees the world. Those with dolichocephalic skulls are not able to focus on objects or individuals as well as dogs with brachycephalic skulls – and short noses. This is because the visual fields of the eyes of brachycephalic dogs overlap far more than dolichocephalic dogs'. This means dogs with broad skulls and short noses have stereoscopic vision that is more like that observed in humans than dogs with longer skulls.

Helton argues that larger dogs tend to be more brachycephalic, meaning they have broad skulls. Broad skulls mean eyes that are more human-like in that they are more forward-facing and have a greater distance between their eyes than is seen in narrow-skulled dogs. This improves vision and depth cues. Helton found that large dogs do better on pointing tasks than small dogs, which may mean that size is the only reason they vary.

As you can see, there is not a lot of research out there, and not a whole lot of agreement within the research, but that is the fun of a scientific revolution. It is supposed to be a messy, opinionated, data-based conversation. The more data you collect, the louder you get to shout, and that is how progress happens.

Morphology Strikes Back

So back to the cleverest breed. Usually, tests assess a particular kind of intelligence – working or obedience intelligence. Ádám Miklósi and his group called this personality trait 'trainability'.

Helton looked at the list and compared the top-ranking elite breeds with actual performance in agility competitions. Agility competitions are a good assessment of trainability, since the dogs must carry out a variety of tasks at the request of their owners.

Agility is made up of two parts: *precision* in following the commands, which is related to the personality trait of trainability, and *speed,* which probably has more to do with a dog's physique. The elite breeds (including Border collies, German shepherds, and retrievers) were definitely faster. Since conventional wisdom assumes that these breeds are the most intelligent, there are more of them competing in agility contests, and therefore, most of the medals belong to these breeds.

In regards to *precision*, however, the elite breeds were about the same as the other, lower-ranking breeds such as Chihuahuas and shih tzus. And some breeds considered to be lower-ranking were actually more precise than the elite breeds. Also, since Helton was assessing trainability, he asked owners how much time it took to train their dogs. If dogs were more trainable, they should take less time to train than a less trainable dog. But in this respect, there was no real difference in terms of the elite breeds and low-ranking breeds.

What Helton did find was that the elite breeds that people *perceive* as being more trainable all look similar. They are all average-looking dogs with medium skulls. They do not have short legs like dachshunds, massive physiques like mastiffs, long skulls like greyhounds, or broad skulls like bulldogs.

Though all these breeds did just as well as the elites in terms of precision, you can imagine a dachshund waddling around an agility course would *look* far less impressive than a fleet-footed Border collie, even if the dachshund was correctly following all of the commands. Maybe when we think of 'smart' breeds, it has less to do with how clever they are and more to do with what they look like. Which just goes to show, don't judge a book by its cover, and don't judge a dog by the shape of her skull.

In talking about breed differences, it all comes down to this: If you think your dog's breed is the best breed, the bad news is, there is no scientific evidence to back you up. But the good news is, there is no evidence to contradict you, either.

217

10

TEACHING GENIUS

How do you train a cognitive dog?

I adopted Milo from a no-kill rescue home just as spring was taking the edge off a frigid Boston winter. He was found wandering the streets with no tag, and in the ten days he had been at the home, no one had reported him missing or come to find him. When I visited the home, the dogs wagged their tails frantically, yipping and yelping, trying to get a hug from anything on two legs. Milo was regal in comparison. Despite his size, he gracefully padded towards me.

'Hey, buddy,' I said, crouching down in front of the wire mesh. His white coat floated around him like a cloud as his tail began to wag. He looked at me serenely, holding my gaze with deep brown eyes that were bright with intelligence. He was the most beautiful dog I had ever seen. He looked like a cross between a Labrador and a polar bear.

I came back for a second visit to decide if Milo was the one. I sat down on the floor. When Milo entered the room, he jumped into my lap – all twenty-seven kilograms of him – and did an awkward but happy wriggle.

It was decided – his was my new soul mate. We were going to do great things. I hoped he might even take up where Oreo had left off and push the boundaries of dognition. Milo taught me more than I could have imagined, but it was not what I was expecting.

The first sign that Milo was 'special' came a week after I took him home. Everything was going great. He was a super-calm dog, and true to the rescue home staff's word, he never peed inside my apartment.

One afternoon, I was sitting with my friends in Boston Common when an ambulance siren went off in the distance. Milo arched his head towards the sky and howled a long, mournful note. My friends and I laughed. How cute. Then Milo started to pant and leaned against my legs. I ruffled his ears, still not getting it. It was a harbinger of things to come.

I wanted to get a dog because I knew I would be living alone in Germany and I thought some unconditional canine love would keep me from becoming lonely. I got more than I bargained for. Milo became the dog who loved too much. For nine months, I lived like a hostage. I took Milo to work every day. I took him to dates at restaurants at night. I even took him with me to the toilet, because if I left him alone for even a few minutes he would howl as though he were dying until I came back. I almost got kicked out of my flat, because the few times I absolutely had to leave Milo, he howled so loudly, my neighbours thought something was wrong and called the police.

Luckily, the Germans are very progressive when it comes to dogs. Dogs are allowed in office buildings, shops, cafés, and restaurants. They are even allowed on buses and trains. The only place they are not allowed is the supermarket. Whenever I needed food, I would have to shop as fast as I could while Milo howled outside.

Oddly, Milo was not very affectionate. He was more like a cat. He would put up with it if I hugged him, but he did not look like he was enjoying it. He never again jumped in my lap. He did not play fetch. In fact, he did not play at all. No tug-of-war, no chase, nothing. Mostly he just hid under any kind of table – the kitchen table, my desk – and slept.

I held off giving him anxiety medication because I thought with time and training he would get better. But no amount of training seemed to help. Milo was disobedient beyond belief. His calm, intelligent demeanour the day I saw him in the sanctuary was a farce. He was not stupid, but he was extremely stubborn. I have successfully trained five dogs, and four of them were rescue dogs. Not one of them had any kind of obedience or behavioural problem. Then there was Milo.

I spent hours a day teaching him basic commands. As soon as we were outside, or anywhere it mattered, I could just forget about it. He would not sit or come when I called him. He certainly would not stay. He could barely walk on a lead. He was twenty-seven kilograms of pure determination, and he wanted to sniff every molecule of dog urine in a five-mile radius. He was so obsessed with other dogs' markings that he completely ignored real dogs. I was a 'dog expert' unable to train my own dog to sit on command.

The Blueberry Factor

I did what anyone would do. I read online advice columns and do-it-yourself training books. I was immediately struck by the underlying message: It was my fault. I had either reared Milo incorrectly or I had not worked hard enough to train him. The universal answer was consistent reward or punishment.

Having just done my PhD on the importance of temperament for co-operation and communication in dogs, I thought this solution seemed too simplistic. How we nurture a dog affects how they behave, but so does their nature. I started to wonder if Milo's temperament made him disobedient and too attached to me. It occurred to me I might be Milo's prisoner for the rest of his life.

The answer was Milo's tongue. Milo's tongue was as blue as a blueberry. When I adopted him, I did not know that only one breed of dog has a blue tongue. Chow chows originated more than two thousand years ago and are one of the nine most genetically wolf-like breeds. Chinese paintings and figurines from the Han dynasty depict chow chows standing proudly as guard dogs or reclining majestically under tables. Chow chows were also occasionally cooked and eaten, and probably got their name from the Cantonese *chow*, 'to stir', as in *chow mein*. Chow chows are also famously attached to and protective of their owners, extremely stubborn, and difficult to train.

I did not have a disobedient dog because I was a failure as a trainer. My dog was genetically more wolf-like and ruled by a different temperament than other dogs. My failure was an interaction between Milo's wolf-like nature and my non-breed-specific training techniques.

I was saved almost by accident. Because of the luxuriousness of Milo's coat, the rescue home staff did not notice that Milo was fully intact until I picked him up to take home. I promised I would have Milo neutered, but in the rush to move to Germany, I never got the chance until nine months later.

When the anaesthetic wore off, Milo was a different dog. The biggest surprise was that with no additional training, he obeyed my commands. The whole time I had been teaching Milo 'sit' and 'stay' and 'come', he would act like he had no idea what I was talking about. Now I knew he had understood the commands perfectly – he just was not obeying them. Suddenly he could walk on a lead. He stopped howling. He was still anxious when I was away, but it was a major improvement.

Milo's temperament changed when his testes were removed, which reduced the androgen levels in his body. As a result, cognitive skills that were there all the time finally had a chance to influence his behaviour.

Milo is the perfect example of how nature and nurture interact to shape behaviour. He also shows how training some animals can be more complex than others, depending on their temperament.

I thought my experience with Milo was an amusing anecdote without much relevance, until I was invited to give a keynote lecture at a dog training conference. After my presentation on the evolution of dog cognition, I was excited to listen to everyone else. I was not disappointed. Many speakers encouraged the use of rigorous methods to analyse the effectiveness of training techniques, while others discussed the role of emotions in driving problematic behaviours. Several pit bulls turned out to be some of the sweetest and best-trained dogs I have ever seen, despite authorities recommending they be put down after the dogs were rescued from a dogfighting ring. There was also canine freestyle, which included everything from a moonwalking Chihuahua to a waltzing poodle. It was clear that these people knew how to train dogs.

However, I was surprised when I realized that a long-rejected school of animal psychology still captured the imagination of many dog trainers. Speakers repeatedly brought up the power of conditioning methods to solve all types of behavioural problems. Trainers advocated consistent rewards to nurture appropriate behaviour. Clickers were out in full force.

It was all fun and games until one of the key speakers began extolling the virtues of behaviourism. Photos of rats and pigeons in Skinner boxes from decades ago flashed across the screen, and we were told that classical and operant conditioning could be used with equal success with dogs, chickens, and every other animal in creation. Next, there was an ode to B.F. Skinner for discovering the universal principles of learning that reportedly revolutionized our understanding of animals.

It was like a spaceship had landed and a whole bunch of aliens had jumped out and announced they were taking us back to the 1950s. Before I go on, I should explain what behaviourism is, how it changed the face of science and society to such an extent that it still lingers today, and how the Skinnerian view of learning was rejected and replaced by a cognitive approach.

The Tyranny of Behaviourism

It is hard to imagine the chokehold behaviourism had on behavioural science for most of the twentieth century. Today, there is a whole range of approaches to studying cognition. There are ethologists, behaviourists, neuroscientists, and people like me who study cognition from an anthropological perspective. But in the US from 1913 to 1960, and perhaps longer, there was only one approach to animal psychology – behaviourism. If you were not a behaviourist, you could not get a job, because all of the hiring faculty were behaviourists. You could not get a grant, because everyone reviewing your grant was a behaviourist. You could not get published, because everyone reviewing your paper was a behaviourist.

Behaviourism started as a reaction against the introspective psychology championed by Freud and others. Their work is too vast to go into here, but if you ever flick through Freud's *Passing of the Oedipus Complex,* you will find passages such as:

The little girl who wants to believe herself her father's partner in love must one day endure a harsh punishment at his hands, and find herself hurled to earth from her cloud-castles.

It was no surprise that some psychologists were looking for a change. They began watching chicks blunder their way through mazes, then slowly get better at finding their way out over time. This was not intellect at work, but something much more elementary.

Learning would replace any need for more abstract ideas about the mental lives of animals. Having focused on a small number of species, behaviourists soon argued that all species learned the same way.

J.B. Watson went so far as to claim that 'no new principle is needed in passing from the unicellular organisms to man'. Study learning in one animal and you understand them all.

The god in the temple of behaviourism was Burrhus Frederic Skinner. If it is hard to imagine the power of behaviourism fifty years ago, it is almost impossible to realize how famous Skinner was as the head of that movement.

Today, Americans rate their top three science role models as Bill Gates, Al Gore, and Einstein. Note that two of these people are not even scientists, and one of them is dead. In 1975, the best-known scientist in the US was Skinner. He was on the cover of *Time* magazine in 1971. He was a guest on the top-ranking daytime TV programme, *The Phil Donahue Show*, and named one of *Esquire*'s '100 Most Important People' in 1970. His novel, *Walden Two*, sold 2.5 million copies and his non-fiction tome of 1971, *Beyond Freedom & Dignity*, was on the *New York Times* bestseller list for twenty-six weeks.

However, you do not make scores of friends without making as many enemies. Skinner in his time was called the 'Darth Vader of

American psychology . . . the Hitler of late twentieth century science itself', 'visibly insane', and a 'fatuous opinionated ass'.

Skinner did not do much to help his image. He was the archetypical boffin, complete with white lab coat and Coke-bottle glasses, fiddling with rats in his laboratory. He was considered conceited even at university and seldom displayed much emotion. His younger brother died of a massive cerebral haemorrhage right in front of him, and the eighteen-year-old Skinner described the entire event with remarkable detachment. During his first TV appearance, he said that he would rather burn his children than his books because 'his contribution to the future would be greater through his work than through his genes'.

Learning Skinnerian

It was at graduate school at Harvard that Skinner discovered behaviourism. Skinner was a fan of Pavlov, who came up with 'classical conditioning'. Pavlov was studying the digestive system of dogs who had an annoying habit of salivating at the sight of their keeper before there was any food in sight. This distorted the data Pavlov was trying to collect, but after a while he realized he was on to something. One explanation was that dogs realized mealtimes were approaching and were salivating in anticipation of the food. This hypothesis assumes that dogs are capable of thought and reasoning. But there was another explanation – the sight of the person who brought the food was a stimulus that caused an automatic physiological response.

Pavlov showed that he could induce a dog to salivate in response to any number of stimuli (buzzer, bell, flashing light). He called this phenomenon conditioning, because the stimulus produced a 'conditioned' response. Over time, it became known as classical conditioning. A perfect example of classical conditioning is clicker training. You click the clicker, the dog looks up at you, and you reward the dog with food. You do it again and again, until every time you click, the dog looks up.

Skinner went one step further, and instead of a clicker or a bell being the stimulus, the animal's behaviour becomes the stimulus. An example of this is when you train your dog to sit. Every time you say 'sit' (the stimulus) and the dog sits, you give him a treat (the response). Over time, the behaviour is reinforced. For the first or second time, you might need to push your dog's bum down at the same time. But after the tenth time, your dog is reliably sitting every time you say 'sit'. Any thoughts, questions, or reasoning going on in your dog's head, Skinner would say, were irrelevant. All that matters is the behaviour you are reinforcing. Skinner called this operant conditioning, because the dog's behaviour is operating on the environment to produce the response or reward.

To train his animals, which were mostly pigeons and rats, instead of a clicker, Skinner built an apparatus called a Skinner box. This is essentially a box with something inside that the animal can manipulate – a lever for a rat or a button for a pigeon – and something that delivers the reinforcement, like food, visuals, or sound. The box is connected to an apparatus that measures how many times the button or lever is pushed.

Using operant conditioning, Skinner could train animals to do amazing things. He had pigeons that could play Ping-Pong and the piano. He trained rats to drop a marble in a hole, which became known as 'rat basketball'. One of Skinner's first projects, during World War II, was called 'Project Pigeon'. The aim was to install in bombs pigeons who were trained to peck keys that would guide the bombs towards their target.

Seeing animals performing these actions for the first time, you might think they were incredibly intelligent, using memory and reasoning. But Skinner argued that training animals is just a matter of stringing together a sequence of actions for which the animal has been rewarded.

To be clear, Skinner did not say that there was no mind, or that animals did not have one. He just did not believe the mind was important, and it certainly did not have any place in psychology. He believed there was no way you could work out what was going on in the mind of an animal. The only thing that counted was the behaviour he was trying to produce.

Life in a Skinner Box

With animals doing exactly what scientists wanted them to, it was only a matter of time before scientists started applying the same principles to people. The disconnect between the scientist and the mental state of another person made Skinner uncomfortable. If you could not directly observe a phenomenon, then as a scientist, you should not count it as data. To gather data on the inner feelings or mind of people, you had to rely on them telling you what was going on. There was no way of knowing if they were telling the truth.

Skinner argued that most of society's problems are not about what people think and feel, but their harmful behaviours. All of a sudden, it was not just pigeons and rats going into Skinner boxes, but humans. In the mid-1950s, Sidney Bijou built a Skinner box for children out of a house trailer that he could tow to schools around Seattle.

In 1976, Douglas Biklen from Syracuse University in New York spent five months observing fifty-three schizophrenic women participating in a behavioural modification programme. Each woman had five desired behaviours assigned to her, such as 'use toilet paper when wiping yourself' and 'stop pounding the floor and walls'. There was also a set of general behaviours, like their personal appearance and housework. They were also expected to participate in songs such as 'London Bridge Is Falling Down and games such as flying paper planes.

The token economy programme, where good behaviour was rewarded with tokens that could be exchanged for privileges, was defined as a success. Over six years, 89 percent of the women participated in at least an hour a day of work. Undesirable behaviours, such as self-starvation and wearing excessive clothing, were reduced.

Soon all sorts of institutions were adopting the token economy. It was applied to children in nursery, violent offenders in prison, the physically and mentally challenged in skills workshops, and juveniles in delinquency centres. In 1969, there were twenty-seven token economy programmes in twenty hospitals, involving more than nine hundred patients.

As for what the people participating in the programme felt or thought, in true behaviourist fashion, this was completely ignored.

One of the sticking points in the public eye was that behavioural modification relied on deprivation. Though Skinner did not find inflicting pain on animals productive, because it produced escape and avoidance behaviour, he did rely on food deprivation. He kept his rats at about 80 percent of their normal weight to keep them motivated enough to do the tests. Prisoners and mental patients were perfect subjects, because they, too, could be deprived, if not of food, then of privileges.

By 1974, the public was responding in alarm to the ethics, or lack of them, in biomedical and behavioural research. The public began to question whether it was ethical to deprive prison or mental inmates of privileges, especially when participation in the programme may not be voluntary. At the time, both behavioural and medical research was being done on the mentally impaired, dying, and prisoners, and this began to look like human rights violations. Soon behaviour modification programmes had their funding withdrawn and were shut down.

The end of these programmes did not stop behaviourism in its tracks. On the contrary, once behaviourism stopped being an institutional tool, it dispersed into the wider community. Behaviourism was used to cure every unwanted behaviour, including weight gain, smoking, speech problems, autism, and irrational fears.

The overall application of behaviourism was limited and the chokehold of behaviourism on the scientific world loosened. Behaviourism depended on four things:

1. Behaviour is dictated by nothing more than a series of stimulus-response mechanisms.
2. In reaction to consistent stimulus, the response should become stronger over time (that was what the charts coming out of the Skinner boxes showed so beautifully).
3. All animals and humans are uniform (what works for a pigeon should work for a dog, pig, rat, human, et cetera).
4. All behaviour can be predicted and controlled, and therefore the inner workings of the mind (thoughts, memories, emotions) are irrelevant.

All of these principles are strictly incorrect, though they have some limited practical value if applied in very particular circumstances.

For instance, conditioning has worked to control phobias. Most dogs will whine and cringe during a thunderstorm. But in rare cases, their fear will develop to such an extent that they will destroy furniture, claw at windows and doors until they bleed, or defecate on the floor. In the most severe cases, a small number of dogs may develop an illness or have a fatal heart attack.

Phobias may be irrational, but they produce a physiological response. The heartbeat increases to get oxygen and blood to the muscles. Sensitivity to pain decreases. The digestive system shuts down, and bodily secretions stop so that the mouth feels dry. The bladder and bowels might evacuate, and pupils get wider to let more light into the eyes.

You may think that you can comfort your dog through a thunderstorm, but one study showed that the presence of an owner did

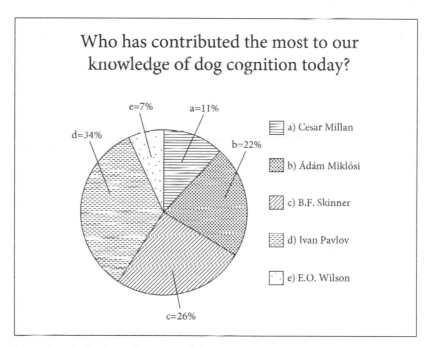

More than half of people surveyed (60 percent), believe either B.F. Skinner or Pavlov contributed most to what we know about dog cognition today. Personally, I think the largest contribution has been from Ádám Miklósi from the Eötvös Loránd University in Budapest, Hungary.

nothing to reduce cortisol (a stress hormone) in dogs (although having another dog around did help).

Treatment for phobias relies on gradual and repeated exposure to the stimulus and, in some cases, receiving a food reward. So in the beginning, dogs may be exposed to a recording of the thunderstorm at the lowest level in a relaxed setting. The dogs learn a positive or neutral association with the noises of thunderstorms, and over time the volume increases until they can get through a full-scale real version without serious effect. This method can be used for various phobias including gunfire, hot-air balloons, fireworks, bees, and high-altitude aircraft.

Nevertheless, these Skinnerian principles are not useful as a basis for understanding and enjoying the company of your dog.

The Cognitive Revolution

It was language that finally did behaviourism in. As we have discussed in relation to Rico and Chaser, human children learn words by making inferences about what sounds refer to which actions or objects – not trial and error.

Skinner attempted to explain how children learn the myriad subtle rules of grammar through stimulus response in his book *Verbal Behavior.* A little-known (at the time) linguist named Noam Chomsky published such a scathing review that Skinner's reputation never quite recovered.

Worse yet, behaviourism could not explain the behaviour of animals. Behaviourism states that animals learn through trial and error and that results should improve over time. As we have seen, animals make inferences, not all animals are capable of making the same inferences, not all animals learn the same things, and not all animals learn the same way. Many of the differences among species are related to the types of problems they must solve in the wild. Animals have evolved a variety of cognitive abilities depending on what they need to survive. All of these inconsistencies finally put an end to Skinner and the school of behaviourism.

By the way, the now legendary Chomsky argued that there must be some kind of innate knowledge that children are born with to allow them to learn a language. The concept of this innate knowledge, which allowed children to learn any language in the world, and its complex rules of grammar, pulled the rug out from under behaviourism. Currently, Chomsky's view is being replaced by volumes of developmental data about how children actually acquire language using cognitive skills only found in humans. The results suggest that humans are not born with an innate universal grammar, but are born with social cognitive abilities that allow them to learn through inferential reasoning and instruction how to use their culture's language.

We now know that there are many types of intelligence. One individual or species can be better or worse than another at solving problems. Also, how good they are at solving one type of problem does not necessarily predict how good they are at solving another type of problem. Just as some people can be geniuses at some things, animals can be geniuses in some areas but not others.

Learning is just one type of intelligence. The cognitive approach celebrates many different types of intelligence and liberates us from the idea that intelligence is a linear scale with sea sponges at the bottom and humans at the top. Asking if a dolphin is cleverer than a crow is like asking if a hammer is better than a saw. Which is the better tool depends on the task at hand or, in the case of animals, which challenges they must regularly confront to survive and reproduce.

Back at the training conference, in a small-group discussion, we started talking about behaviourism. Many trainers said their success relied on behaviourist techniques such as clicker training and positive reinforcement. They seemed to believe they were using operant and classical conditioning to train dogs. But when they talked about *why* the techniques worked, they put the dog in a cognitive framework. They spoke about how 'the dog *knows*', or 'the dog *wants* to perform' – yet true behaviourists reject thoughts and desires as irrelevant.

A cognitive approach works so well with dogs, not because they have no mind, but precisely because they do. When dogs solve a problem, they can make inferences or generalize what they have learnt to solve a new problem. For example, each new person who asks a dog to sit does not need to teach the dog to sit. If you train dogs to sit, they will not only sit for you, they will sit for a new person in a different location. Similarly, if you point in the direction you threw a ball, dogs do not go looking for food. They understand during a game of fetch that your point refers to the ball and that during a food-finding game your point refers to the food. Pointing gestures mean different things in different contexts. Dogs can make these types of generalizations because they are cognitive.

The genius of dogs is their ability to understand human communication and their motivation to co-operate with us. Their genius is probably why they are so easy to train. But dogs also have biases and limitations to their understanding of how the world works. A cognitive approach allows us to train around these biases and limitations instead of fighting a losing battle against them.

Dognition relative to cognitive ablity in other mammals

Cognitive ability	Vapid	Similar to other mammals	Remarkable	Genius
Comprehending visual gestures				X
Learning new words				X
'Talking' through vocalizations and visual signs			X	
Understanding an audience's perspective			X	
Navigating in space		X		
Individual learning/associative learning (conditioning)		X		
Understanding physics	X			
Quantity judgements (counting)		X		
Self-understanding		X		
Learning from others		X		
Copying others' actions			X	
Recruiting help			X	
Detecting cheaters		X		
Empathizing		X		
Feeling guilt		X		

Current Schools of Training

There is very little published research on different training techniques. While certain techniques may work with certain dogs, these techniques are not based on science. A cognitive approach can help us better understand how dogs think, which in turn can help us develop more effective training techniques. Hopefully, we can transform the art of dog training into a science. I think both ends of the lead will benefit.

Currently, there are two main approaches to dog training: the 'top dog' school and the 'more is better' school. The top dog school suggests that owners should establish a dominance relationship over their dogs in order to ensure they are obedient. It originates from the idea that wolf packs have strict dominance hierarchies, where the wolves compete for dominance but are held in check by the alpha male and female. Since dogs evolved from wolves, the top dog school encourages you to act like the alpha wolf. This can include anything from using choke collars and never letting your dog walk first through a doorway, to the 'alpha roll', where you flip dogs onto their back and hold them by the throat.

The problem from a scientific perspective with the 'dog in wolf's clothing' approach is that it assumes that the social system of dogs is the same as wolves'. However, domestication has changed the social system of dogs. A comparison of feral dogs and wolves reveals a number of important differences in their social structure, as we saw in Part Two.

The most relevant is the relaxed hierarchy in feral dogs. Leaders in a feral dog pack are not the most physically dominant individuals. Instead, dogs with the strongest affiliative bonds or friendships in the group are the most likely to be the leaders.

Some supporters of the top dog school have suggested that how you play with your dog can affect how they view your dominance relationship. For instance, you should not let your dog win a tug-of-war match, because your dog will think they are dominant over you.

In one of the few experiments on training techniques, researchers assessed how a group of golden retrievers reacted when someone took their food or toys away and how long it took the dogs to obey commands both before and after they either won twenty games of tug-of-war or lost twenty games of tug-of-war. Regardless of whether the golden retrievers won or lost the games, they did not show an increase or decrease in dominance towards their human partners. This suggests we do not need to dominate or stop playing with our dogs to improve their training.

The 'more is better' school is influenced by the behaviourists' linking stimulus and response to shape behaviour. The little research that has been done does not suggest that maximum reward or training is the most effective technique to shape a dog's behaviour.

In humans, rewarding someone for a behaviour actually reduces motivation once the reward is reduced or taken away. For instance, say a child enjoys reading, and then you start rewarding her with chocolate for reading. Once you stop rewarding her with chocolate, she is unlikely to enjoy reading for its own sake. This is called the 'overjustification effect'.

This effect seems to be present in dogs. In an experiment, dogs were rewarded with average food for obeying a command. The average reward was then replaced with a better reward. When the researchers switched back to the average reward, the dogs' performance dropped. This means rewarding dogs with delicious treats does not necessarily lead to faster learning. It also means you will not be able to switch back to ordinary treats if you run out of the good stuff. Rewarding dogs with food for doing something that they already do for praise could have the same effect.

Another theory of the more is better school is that the fastest way to train dogs is long, repetitive training sessions each day. Recent research suggests we might be able to relax a bit. In one study, dogs were trained to put their front paw on a computer mouse pad – something they had never done before. The dogs were either trained five times a week or once a week. Dogs who had only one training session a week learned in fewer sessions.

In another study, dogs were trained to go to a basket and stay. Their training sessions varied in frequency (one to two times per week versus every day) and duration (one training session versus three in a row). Dogs learned best when they were trained once a week in a single session. The worst performance was in dogs receiving the most training – every day for three sessions. These results show that when it comes to learning, sometimes less is more.

'Clicker training' is one of the most popular training techniques today. Clicker training is based on classical conditioning as discovered by Pavlov and perfected by Skinner and the behaviourists with rats and pigeons. A clicker is a metal tab that makes a clicking sound when you press it. In response to a desirable behaviour, the click functions as a 'secondary re-inforcer' when paired with a reward. Clicker enthusiasts believe the clicker is effective because you can click almost as soon as you see the behaviour you want, as opposed to rummaging round for a treat, which might cause a delay. After the behaviour has been 'marked' by the click, the reward can quickly follow.

Given the clicker's popularity, it is surprising that there is only one study that compares the clicker with other types of training. Basenjis were trained to touch their nose to a cone in response to either a click and a food reward, or just a food reward. The group of dogs trained with the clickers in addition to food did not learn any faster than dogs who were rewarded only with food. At least for the moment, there is no scientific evidence to support the theory that clicker training facilitates faster learning in dogs.

Clickers do help, but we need to determine exactly when they help and why. I suspect that rather than helping dogs learn faster, clickers make people better trainers. Using a clicker might help owners reward their dogs more consistently or even make owners feel as though they have more control during training. We need more research to know the answer.

Current training methods do work, since dogs can be trained to do so many amazing things. But from a scientific perspective, we do not know which methods work best. There is no formalized training that combines what we know about dog behaviour and training with the latest research in dognition.

Cognitive training would not only identify the different ways that dogs learn but also identify limitations and biases that can prevent learning. Strategies can then be designed to work around these biases and limitations while tapping into the genius of dogs.

Dognition Working *for* You

The most important lesson about dognition is that when dogs are left to their own devices, they are completely unremarkable. Imagine you were placed in an empty room and every time you stood near the door, pound notes slid underneath. You would quickly recognize the relationship between your action and its result.

In comparison with wolves, dogs are slower at forming similar associations between arbitrary cues and the presence of food. We have seen that wolves learn and unlearn which hiding location to search under based on its colour much faster than dogs. Wolves are also much faster to learn how to make a detour around a physical barrier than dogs. When dogs are on their own and have to rely on trial and error to form an association, the results are not very good.

Wolves might be better than dogs at trial-and-error learning, but no one would argue that wolves are easier to train than dogs. So trial-and-error learning cannot be the unusual skill in dogs that makes them so trainable.

The trainability of dogs relies on other forms of cognition. A dog will always learn from a human faster than a wolf, because dogs have evolved skills to read our communicative signals. While working dogs might be more skilled at using human gestures as a result of either training or human selection for this skill, all dogs are skilled at using human gestures. Even rescue dogs and breeds not intentionally bred by humans are skilled at using human gestures.

A cognitive approach also helps us identify the context in which dogs are most likely to learn from our attempts to communicate with them. For example, dogs are more skilled at using our gestures if we pay attention to them while giving the gesture. Like infants, dogs are best at following the direction of your gaze when you signal the communicative nature of your head movement. Dogs are more likely to look where you are looking if you call their name and make eye contact before shifting your gaze.

Dogs are less skilled at using a gesture that is not intended as communicative. If you extend your arm as if you are pointing but then look at your watch, dogs are less likely to use your pointing gesture. Dogs have trouble understanding threatening gestures to prevent them from going somewhere.

Dogs are more likely to follow your pointing gesture if you use a high-pitched voice (not even necessarily their name) to attract their attention before gesturing. Dogs are also more determined to search for a hidden object when you use a high-pitched voice as opposed to a low-pitched voice.

Making eye contact with your dog, calling her name, and encouraging her with a high-pitched voice will maximize the likelihood that she will use your gestures.

The success of verbal signals varies. In some cases, verbal signals may help dogs learn faster, while in some cases it may confuse them. When dogs are learning to solve a new problem, several studies have shown that talking while demonstrating the solution helps dogs pay attention and learn from the demonstration.

However, the same approach can have a negative effect. If you say a series of words that includes the command 'sit', dogs are less likely to obey than if you said only the word 'sit'. This effect is strongest when dogs are asked to respond to a new command or a known command in a new location. This suggests you should say only the command you want, rather than chattering.

Cognitive training will develop techniques that rely on more flexible or faster forms of learning, rather than trial and error. One example is the ability of dogs to 'learn how to learn'. Learning how to learn allows a dog to generalize a newly learned skill to a new situation without having to start from scratch. Sarah Marshall-Pescini of the University of Milan compared highly trained working dogs with untrained pet dogs. The dogs had to find and use the lever that opened a box with food inside. While the untrained pet dogs quickly gave up on trying to solve the problem on their own and simply looked to their owners, the trained dogs persisted until they found a solution. Marshall-Pescini suggests that the trained dogs may have learned to learn how to solve such new problems in their everyday lives as working dogs.

Another lesson from dognition is that dogs can almost instantly learn to solve various problems by watching a demonstration. Puppies were kept with their mother while she was on the job searching for drugs. When these puppies then went through drug detection training, they were four times more likely to get the highest training scores than other puppies.

Also, while dogs struggle with problems such as opening boxes and doors and finding their way around barriers, these problems are easy for dogs to solve if they see someone else solve the problem first.

Dognition Working *against* You

Sometimes, dognition can make dogs too clever to obey us. A cognitive dog is not programmed to obey your every command. In conflicts of interest, we can expect our best friends to use their intelligence

to manoeuvre around us. For example, remember that dogs sometimes know what you can and cannot see. They are more likely to
drop a ball in front of you than behind you, to beg for food from
someone who can see them, or to fetch a ball that a human can see.

Christine Schwab and Ludwig Huber from the University of
Vienna wanted to find out how dogs decide when it is safest to disobey
their owners. In the experiment, the owner told the dog to lie down,
then put the dog's favourite food in front of him, about a metre and a
half away. The owner then stood so that the dog was between the
owner and the food. Then the owner did one of these things:

- Looked at the dog: The owner sat in a chair so her eyes,
  head, and body were turned towards the dog.
- Read a book: The owner sat in a chair with her head
  and body towards the dog, but her eyes were absorbed
  in a book.
- Watched TV: The owner's body was turned towards
  the dog, but her head and eyes were turned away from
  the dog towards the television.
- Turned around: The owner sat in the chair reading a
  book with her back turned towards the dog.
- Left the room: As soon as the owner put the food in
  front of the dog, she left the room and closed the door.

First, anyone without a perfectly behaved dog can feel better – 60
percent of the time, the dogs could not resist eating the food no matter where the owner was facing. The rest of the time, if you had to
guess under which condition dogs would most likely disobey the

command and snatch up the food, you would probably guess it was when the owner left the room – and you would be right.

The next most likely condition where dogs would sneak up and eat the food was when the owner's back was turned, showing that dogs understand the difference between a human's front and back, and that a back turned towards them means the owner is not paying much attention.

This has been replicated many times, including our study where dogs prefer to drop a ball in front of you where you can see it, rather than behind you. Similarly, dogs were also less likely to obey when the owner was watching television and the eyes and head were turned away from the dog.

Perhaps this just means that dogs have learned that the more of you that is turned towards them, the more diligent they should be in obeying you.

However, dogs were more likely to disobey the command and eat the food if the owner was reading a book than if they were looking at the dog. This is impressive because the two conditions were almost identical – the owner was sitting in a chair with their body and head facing towards the dog.

The only difference was that in the reading condition, the owner's eyes were facing downwards. Eye contact was what made the dogs obey the command most often for the longest time, showing that they really do understand what their owners can see, and they use this knowledge to decide how obediently to behave. Dogs are not above using their intelligence to make sure they get what they want.

Whenever I said a command to Milo inside the house, he would obey me perfectly, but as soon as we got outside, he would act as though he did not know what I was talking about. Researchers from De Montfort University and the University of Lincoln found that when owners stood at a distance of two and a half metres from their dogs, their dogs were less likely to follow a 'sit' command than if they were standing right in front of their dogs. Dogs were even worse at following the command if their owner was hidden from sight.

So when your dog is refusing to obey you, keep in mind that while they want to please you, they are not above trying to figure out when they can get away with ignoring you. A happy relationship with your dog may literally depend on your ability to keep an eye on them.

Cognitive training will also take into account that different cognitive skills can interfere with and constrain learning. For instance, dogs can have cognitive biases that prevent them from seeing the answer to a problem even when it is right in front of their eyes. Remember from Part Two that if a doorway is moved to another location, dogs will still try to exit from the old doorway even though they can clearly see the new doorway and the solid wall that has replaced the old one. Dogs' memory of the old doorway interferes with their ability to rely on what they can see.

Another bias involves the scenario where dogs respond to a human's gesture instead of what they have seen with their own eyes. When an experimenter shows where food has been hidden but then points at another location, dogs do not search for the food they saw but instead go to where the human pointed.

Dogs have evolved a bias to pay attention to human communicative gestures even when they contradict what they have just seen. While this can be advantageous, it can also have serious consequences. Bomb-detecting dogs who rely on their handler's behaviour instead of their sense of smell might miss their target.

Cognitive trainers will also recognize that some behaviours are difficult to train because dogs may have laterality biases. For example, some research suggests that some dogs favour one paw over the other for manipulating objects, with females favouring their right paw and males favouring their left. In addition, when dogs encounter emotional stimuli it tends to be processed in the right hemisphere of their brain, leading them to turn their head to the left in response. This might make dogs more likely to move to the left when aroused even if that is the 'wrong' direction for training purposes.

Cognitive training recognizes that your dog is neither a black box nor a furry human. While dogs have evolved a specific type of genius, they have limitations like any other species. Understanding these limitations will enhance training techniques. Just as we do not expect infants to understand certain problems, such as the danger of stairs and knives, the same goes for our dogs.

For instance, contrary to popular belief, there is no experimental evidence that dogs experience the feeling of guilt or that they have a human-like concept of guilt. Currently we only have evidence that dogs react to their owner's frustrated behaviour. This means that trying to train a dog after the fact will not work. After agility training, researchers saw some contestants scolding or even physically pushing their dogs. The only effect this had was to raise the stress levels of the dogs. It is very unlikely that the dogs understood they should feel guilty for their poor performance or that verbal or physical punishment improved later performances. Likewise, if you come home and discover your adorable new puppies have chewed the sofa, tipped the rubbish bin, or had an accident, it is unlikely they understand why you are unhappy. So it is no good pointing and yelling at the sofa stuffing, scattered litter, or mummified poop.

Another cognitive limitation is that dogs do not understand what someone knows or does not know. For example, dogs used showing behaviour (gazing and barking) to help a human find a hidden object, regardless of whether the human had or had not seen the object hidden. This suggests that when dogs are communicating with us, they are usually making requests for things they want. They probably cannot communicate with us based on what we have seen or have not seen in the past.

For hundreds of years, there have been stories of heroic Lassie-type dogs (and at least one kangaroo) running to fetch help when their owners are in trouble. To fetch help, Lassie needs to understand that she needs to inform other people of the emergency because only she saw it. This may not come as a surprise, but based on the first generation of experiments, it is unlikely your dog, no matter how much of a dog genius he is, has the cognitive abilities to do this. Dogs might have only a narrow understanding of different types of threats towards humans. While most dogs understand the threat of strangers, their limited understanding of physics, as we have seen, makes them pretty useless in other situations.

Krista Macpherson and William Roberts from the University of Western Ontario in Canada decided to test how much dogs understood when their owners were in danger. In one test, the researchers simulated a large bookcase falling on a dog's owner. While the heavy bookcase pinned the owner to the ground, the dogs did not seek help from a bystander. The most they did was roam around the bystander's vicinity, even when the owner was crying in pain and calling for help. The dogs did not seem to understand the physics behind the situation, so they did not appear alarmed or motivated to get help from another human.

This of course does not mean that dogs do not save lives – they save lives all the time. For instance, in January 2007, Mike Hambling was crossing a frozen river with his German shepherd, Freddie. Freddie seemed hesitant to cross the river, tugging on his lead in protest. Suddenly the ice broke under Mike, and he plunged into the freezing water. The weight of his clothes dragged him down and he struggled helplessly, unable to pull himself out. He was starting to black out from hypothermia when he felt a tug on his wrist. Freddie was pulling on the lead with all his might. Eventually Freddie dragged Mike out of the ice water to shore. The Purina Animal Hall of Fame abounds with stories like these of animals fetching help in an emergency, and 83 percent of the inductees are dogs. But how much do dogs understand? Did Freddie know his owner was in danger and make the connection between the freezing water and the urgency of pulling Mike towards the shore? Or did Freddie just feel himself being pulled towards the hole in the ice and struggle backwards until he was no longer being pulled?

Dogs can sniff out cancer and get help for house fires or alert their owners to dangerously low insulin levels. Just because your dog is not so good at physics does not mean your dog would not get help for you if something happened. In some situations they might be able to help you. Cognitive training makes us realistic about the ways in which dogs will be able to help us and when they will not.

Cognitive training acknowledges all of the different ways that dogs can learn, whether it is by making inferences like Rico and Chaser, paying attention to our communicative gestures, learning to learn, or solving a problem by watching someone else do it first.

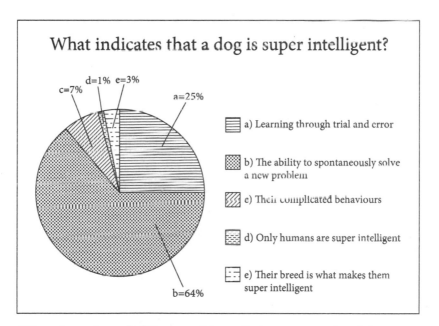

Although most people (64 percent) know that spontaneously solving a new problem is the key to dog genius, one in four people still believe that trial-and-error learning is the secret to a dog's success.

Cognitive training also recognizes that these same abilities can make a dog too clever to obey us. Finally, there are problems that some or all dogs will not be able to solve regardless of how much training they receive.

Milo's Achilles' heel was his barking. Excessive barking is the most commonly reported problem behaviour in dogs. It is natural for dogs to bark, and over the course of their domestication, they seem to have developed a knack for it. Dogs bark more than wolves and in a wider variety of contexts. A bark can mean a greeting, a desire to play, a threat, distress, or just that everyone else in the neighbourhood is barking.

We all know (or have owned) a dog who will not stop barking, leading to antagonism from sleepless neighbours and helpless frustration from owners. Sometimes, you can train dogs not to bark by addressing the context in which barking occurs – for instance, when someone rings the doorbell or knocks on the door. Positively reinforcing the dogs to go to their bed, or a rug, or another spot away from the door and lie there for one minute on command can reduce barking by up to 90 percent.

Another solution that has been relatively well studied is the citronella spray collar. When the dog barks, the sound triggers the collar to release a cloud of citronella with a hiss. The collars seem to work best when dogs wear them intermittently (every second day for thirty minutes) for a period of three weeks. This is surprising; you might think the collar would work best if it sprayed dogs *every* time they barked, but here again, more repetition is not always best. The citronella collar also seems to work just as well as an electronic shock collar in reducing barking when the two collar types were directly compared. However, the citronella collar does not stop dogs from barking altogether, and dogs slowly habituate, with barking levels increasing over time. The week after dogs stop wearing the collar, the barking rate increases, although not to the same levels as before (this also occurs with shock collars).

For chronic barkers like Milo, these solutions are unlikely to help. Excessive barking is usually a product of temperament, and with Milo, his anxiety was the problem. Dogs have varying levels of hormones and neurotransmitters that make them more or less anxious. Anxiety can cause various behaviours ranging from Milo's howling to house soiling. It also affects a dog's cognition. Milo's anxiety affected his ability to learn and perform commands. Eventually, I just got lucky, and Milo happened to be one of those special cases where neutering him relieved most of his anxiety.

Once we overcame Milo's anxiety, I never had another problem. He lived out his days with me in Germany, and the Germans never tired of telling me what a beautiful polar bear I had. My Milo died in 2007. He was looking a little slow and was having trouble climbing the five flights of stairs to our apartment. I took him to the vet, expecting arthritis. Instead the vet found a tumour the size of a basketball inside his abdomen. I was as sad when Milo died as I was for Oreo, even though two more different dogs probably have never existed. In the end, Milo really was the dog I had been hoping for.

11

FOR THE LOVE OF DOG

Could we love one another more?

In rich countries, like the US and the UK, we can afford to spend serious amounts of cash and time on our pets. In other places and cultures, it is different.

When our Congolese friend Gisele Yangala first saw our dog Tassie, she called out cheerily, 'Hey, *tshibela-bela*.' My wife, Vanessa, preened, because in many languages, *bella* means 'beautiful'. Tassie was used to admiration. We call him the Kim Kardashian dog because like many socialites, he is famous for exactly nothing.

Whenever reporters come to do a story on the Duke Canine Cognition Center, Vanessa cunningly makes sure Tassie is always there at the right time. She gives him baths that take an hour and include a conditioning treatment for his fur. So far, he has been in *Time* magazine, *National Geographic*, and on CBC, NBC, and CNN, just to name a few.

I don't want to give the impression that Tassie is gifted – he is not. In fact, when Randi Kaye from CNN TV's *Anderson Cooper 360°* came, she gleefully informed me that Tassie was dumber than a lemur (he had failed a test that she had seen the lemurs at the Duke Lemur Center pass with flying colours). But, as we have seen in modern pop culture, you don't need brains to get famous, and Tassie knocks over the experimental cups quite magnificently and will do it from every angle until the treats run out.

So when the word *bella* was thrown at Tassie, he stood up a little bit taller and let the sun shimmer on his fur. Then Gisele proceeded to inform us that *tshibela-bela* was a Congolese dish of dog meat that is sometimes served in her native town of Lubumbashi.

'You take the dog and stick a corncob in both ends. Then you roast slowly, turning the spit, like you would a goat.'

Vanessa listened, horrified, before hurriedly shooing Tassie out the door, where he would be safe from those corn cobs.

Cultural Differences

Relationships with dogs vary historically and around the world. The skeletal remains of an unusually old dog were unearthed in Anderson, Tennessee, with evidence of multiple maladies including arthritic development due to a chronic infection stemming from an unhealed rib fracture. This dog would only have survived into old age with such injuries if humans cared for its well-being more than 7,000 years ago. The prevalence of such dogs has led archaeologists like Darcy Morey to conclude that

> Clearly, several thousand years ago, as well as in historic times, traumatized elderly dogs were cared for by people in North America, and then buried affectionately when they died.

Likewise, archaeological evidence suggests that before adopting agriculture the Mississippian society revered dogs, since they were commonly buried together with people. This suddenly changed with the advent of agriculture. The rate of dog burials decreased, while the number of cut marks on dog bones increased – suggesting dogs were being eaten once an agricultural economy took root.

The first time I realized that not everyone thinks of dogs in the same way was when I went to the Galapagos Islands. Everything on the Galapagos Islands was like a pet. There are no predators, so the blue-footed boobies, seals, tortoises, and all of the other species were so docile that you could walk right up to them and pat them on the head. But when dogs arrived, they became predators. In particular, dogs eat the Galapagos marine iguanas, the only marine iguanas in the world. In the early morning before dawn, the islanders, park rangers, and tour guides saw dogs slinking across the road and said disgustedly, 'Dogs. We'll have to come back and shoot them.'

I was shocked that dogs could be anything but beloved. Since then, I have seen dogs all over the world, from Uganda, to Australia, to Russia, to Italy.

Raymond Coppinger noted that in Kenya, babies are given a puppy to lick their bottoms clean. Coppinger also visited Pemba Island off the Tanzanian coast. Pembans do not pet their dogs or feed them. There are no fences for these dogs, no chains or collars. In fact, the Pembans believe that God does not like dogs and that if a dog enters their house, the house must be spiritually sanitized before God will visit them. The noses of dogs are cold and wet because evil organisms live inside, and coming into contact with dog drool will probably make you sick.

In Dominica in the Caribbean, there are thousands of stray dogs who, among other things, carry diseases harmful to human health.

Only 12 percent of dogs are vaccinated, and more than half of dog owners take their dogs to the vet only when they are sick. Most pet dogs are confined to the house and yard, while the rest are allowed to roam the streets with strays, scavenging from rubbish bins or back gardens. Not surprisingly, most dogs do not reach old age.

Traditionally, dogs were more important to the Japanese after they were dead, and dogs have always been buried with much ceremony and ritual. There are more than eighty pet cemeteries in Tokyo alone. Samurai gave their dead dogs Buddhist names and had them buried in temples. Ordinary Japanese had their dogs buried in pet cemeteries that were tended by the community to protect them from the spirits of dead dogs who might return to wreak havoc on the living.

Today, dogs can, for a high price, have their own grave, or they can have their ashes stored together with those of other dead pets. The altars of these graves are covered in pet food, toys, and leads. In summer and autumn, there is a ceremony for the pets in the cemetery where the priest calls out the names of owners and their pets for around half an hour.

But the most complex relationship between humans and dogs probably exists in China. On 9 August 2006, *The New York Times* reported on a boy and his father who were forced to march their two German shepherds to a public square in Yunnan and hang them from a tree. It was part of a brutal killing of more than 50,000 dogs in the province. Dogs were dragged from their homes, confiscated while walking with their owners, and clubbed to death on the spot.

The cull was in response to a growing rabies epidemic in China. In the UK and the US, we are largely unfamiliar with rabies, which rarely occurs in domestic animals because of rigorous vaccination programmes. Rabies is a virus in saliva that spreads through a bite or an open wound. Initial symptoms are flu-like and include headache and fever. It can take up to two years for the virus to take hold of the brain, but once more serious symptoms appear, such as a fear of water and wind, drooling, and convulsions, rabies is incurable and death is inevitable.

China has the second highest rate of illness and death from rabies in the world, and up to 95 percent of rabies is transmitted by dog bites. There are up to 200 million dogs in China. In rural areas, where the disease is most likely to spread, poverty prevents people from vaccinating their dogs and treating themselves. In one province, 89 percent of people who contracted rabies were not even given the appropriate drugs. Still, the Chinese government's decision to periodically bludgeon thousands of dogs to death was met with criticism from dog lovers around the world.

In addition, eating dogs still occurs in several parts of China, and photographs of dogs in cages being transported to abattoirs have not helped China's image as an animal-loving country. Dogs have been a part of Chinese cuisine for hundreds, if not thousands, of years. In ancient China, dogs and pigs were the main source of animal protein. Everyone ate dog meat, from emperors to students studying for exams.

At the same time, the Chinese were among the world's first dog lovers. As far back as the twelfth century BCE, there is the story of Emperor Woo receiving a famous dog called Ao from the wild Leu tribe in the west. The dog was a bloodhound who was said to be able to read minds. Then there were fighting dogs, said to be as big as oxen, who enjoyed such celebrity that when they walked through the streets, people knelt down as they passed.

Whereas in Europe, pet keeping did not appear until the Middle Ages, pet dogs in China probably appeared in the first century BCE. Called 'Pai', these small, short-legged, short-headed dogs could fit under the low tables that are traditional in Chinese homes.

From this time on, many Chinese emperors kept small dogs. Though these dogs resembled shih tzus, Pekingese, and pugs, these breeds as we know them did not exist in ancient China. Instead of issuing a breed standard, whoever was emperor decided which was his favourite dog and had them painted in an imperial book. The breeders of the time were palace eunuchs who would try to breed dogs who looked like the emperor's favourite. The highest compliment a breeder could receive was that his dogs could go 'in the book'.

Colours and markings were of great importance – if you had a dog who had a marking in the shape of a phoenix, you could sell it to the emperor for a fortune. A dog with a black coat and a white head was sure to get you an official appointment, because these markings meant the owner would have many sons. A white dog with a black head would make you rich, and so on.

One of the most famous dogs in Chinese history was from the eighth century. The emperor was playing chess with a prince and losing badly. His favourite concubine, legendary for her beauty, was watching discreetly in the distance with her little dog called Wo. Seeing her master losing face, she let Wo run onto the chessboard and knock over all of the pieces. The emperor was delighted.

The cult of lapdogs reached its peak in the nineteenth century. It is around this time that Pekingese dogs started to appear in paintings and on porcelain. These and other dwarf breeds were called 'sleeve dogs', because they could fit in the large, open sleeves that were the fashion.

By the twentieth century, dogs were effectively banned in China. The Communist Party took control of the country in 1949 and declared that keeping dogs was a bourgeois pastime. In 1952, during the Korean War, China accused the US of using biological warfare; they were convinced that the US had released dogs into China carrying lethal infections. Teams of executioners were set up in every city to destroy all dogs. In 1983, just as dogs were beginning to appear again, they were banned from the city of Beijing.

Today, dog ownership in China is complicated. With China's rising prosperity, dogs are once again in fashion – although now it is not emperors who are setting the trend for canine companions, but the rich and elite. The most desirable breed is far from the short-legged, short-faced Pekingese. Instead, the enormous Tibetan mastiff is the prize of the day. Fierce and independent, weighing up to eighty kilograms, the Tibetan mastiff is originally from the Himalayan plateau. Legend has it that Genghis Khan organized 30,000 of them into a canine army to conquer western Europe.

In 2011, a Tibetan mastiff became the world's most expensive dog. A Chinese coal baron paid ten million yuan (one million pounds) for the one-metre-tall eighty-kilogram 'Hong Dong' (Big Splash), whose diet includes abalone and sea cucumber.

Dogs in major Chinese cities now enjoy all the luxuries of their Western counterparts, including hotels, social networks, and swimming pools. To the horror of some, owners are even grooming their dogs to look like exotic animals, including pandas and tigers. As an affluent dog owner, you can take your pooch to a dog-friendly cinema or enjoy a drink at a doggy bar downtown.

On the other hand, rabies outbreaks still lead to the periodic slaughter of tens of thousands of dogs, and lorries full of dogs can still be seen on their way to abattoirs to be sold to restaurants.

The Awful Truth Back Home

I regret to report that the West's treatment of dogs may not be much more humane than the Chinese. The Humane Society of the US (HSUS) estimates that up to eight million cats and dogs end up in the nation's rescue homes each year, and of these, around half are euthanized. The reason most often given for relinquishing a pet is either relocating to another city or a landlord forbidding pets on the property. But when you look more closely at the data, the people who gave moving as the reason for giving up their dog usually also reported at least one behavioural problem. More than half reported that their dog was hyperactive, 40 percent noted that their dog was too noisy, a third said that their dog damaged things inside or outside the house, and 26 percent reported that their dog soiled the house. Furthermore, only 6 percent of the dogs who were relinquished had attended obedience classes or some kind of professional training.

Only 21 percent of people adopt dogs from a rehoming centre. The rest are obtained from other sources, including amateur breeders and puppy farms – the dark underbelly of the purebred world.

Breeding dogs is expensive and time consuming. Responsible breeders have to ensure that all of the dogs have their vaccinations and proper veterinary care. The history of the parent dogs has to be carefully documented to avoid any genetic defects, and good breeders make sure their dogs are not bred too often. All puppies have to be properly socialized to avoid later behavioural problems.

By definition, a puppy farm is a business where profit is considered far more important than the welfare of the dogs. As Wayne Pacelle, president of the HSUS, notes in his book *The Bond,* to maximize profits, puppy farm dogs often live in horrifying conditions. The dogs are kept in wire cages to minimize waste clean-up, and without a solid surface, dogs often suffer from injuries to their paws.

To minimize the noise from barking dogs, breeders can cut dogs' vocal cords by crushing them with a steel pipe. Any cosmetic operations, such as tail docking and ear clipping, are done without a veterinarian or anaesthesia. Parent dogs are bred non-stop from the age of six months to five or six years of age. When the females have been bred to exhaustion, they are either killed or shipped off to rescue homes for possible adoption – if they are lucky.

From an account of a rescue worker who legally raided a puppy farm:

> Kennels have plenty of dogs in them but no food and water. They are filthy from one end to another; the concrete is covered with excrement. . . . There are dead dogs here, some only skeletons, some so badly decayed that only hair and skeletal forms remain. . . . Most of the dogs have missing pieces of ears, eaten away by flies. . . . At this mill a mother dog is found with a litter of pups. The windows and doors are shut, there is no water, and it is 98 degrees. Two of the pups are dead.

The puppy farming business got going after World War II, when crop failure in the Midwest led the US Department of Agriculture (USDA) to encourage farmers to raise puppies. Pet shops began opening everywhere, becoming the main supplier of puppies to the American public, a tradition that has continued for the last half century and become the norm in many countries. The HSUS estimates that there are now at least 10,000 puppy farms in the US, including licensed and unlicensed facilities. These puppy farms produce between two million and four million dogs each year. They also report that almost all pet shops in the US sell puppies from puppy farms, since this is the only way they can keep different breeds in stock. Puppy farms are dotted across the UK, and at least 50,000 dogs are imported each year from puppy farms in Ireland to meet the demand for pets.

Not surprisingly, dogs from puppy farms often have health and behavioural problems. Even if pet shops claim that a dog comes with a health certificate from a vet, these checks are brief and do not test for any of the genetic defects, diseases, or parasites common in puppy farm dogs. For instance, luxating patella, a disease caused by overbreeding the mother, can lead to puppies developing a limp that can cost $3,200 (£2000) to repair through surgery. Also, any health guarantee probably has a time limit, and problems can crop up months or even years after the puppy is bought.

If there is little regulation for pet shops when it comes to puppy farms, there is no regulation at all on the Internet. Puppy farms that sell directly to the public via the Internet have no need for a licence and will never be inspected. According to the American Society for the Prevention of Cruelty to Animals (ASPCA), as many people are buying dogs over the Internet as from pet shops, and 89 percent of 'breeders' selling over the web are unlicensed by the USDA. In 2007, the Internet Crime Complaint Center reported seven hundred cases involving dogs sold on the Internet, including the Nigerian pet scam where puppies are offered for free or at discounted prices, but with many miscellaneous fees. There is also the bait and switch, where the photo of the puppy and the actual puppy people receive are different.

The bottom line is that no responsible breeder will sell puppies to a pet shop or through an Internet site, because they want to meet potential owners in person and make sure they are a good fit. The only way to guarantee puppies are not from a puppy farm is to pick them up from a rehoming centre or a breeder yourself. If a breeder will not let you visit his site, or if he disappears to get the dogs, or you are not allowed to meet the parents, this is bad news. If he sells more than one breed and has more than one litter at a time, this is also a red flag. The dog will be part of your family for at least ten years. It is worth doing some research, meeting the dogs in person, making sure they are happy, healthy, and come from a good home. Alternatively, you can adopt instead of shopping and save one of the millions of rescue dogs or cats who are euthanized each year.

As well as legal abuse of dogs, there is plenty of illegal abuse, one of the most vicious forms being dogfighting. Dogfighting dates back to at least the fifth century BCE, with scenes of dogs in blood sports being depicted by the ancient Greeks and the Libyans. It enjoyed a long tradition in England, where it replaced bull- and bearbaiting, as dogs were easier to get than a live bear. When dogfighting was banned, dogs were still fought in sheds without too much attention; the practice carried on without much interference from the law.

Dogfighting was banned in the UK in 1835, making it the first country to criminalize the sport, and the US followed suit in 1874, though it was not outlawed in all fifty US states until 1976. Today, the sport of dogfighting is far from an occasional illicit pastime in backyard sheds. It has reached epidemic proportions.

There are more than 40,000 dogfighters in US cities, and the Royal Society for the Prevention of Cruelty to Animals (RSPCA) receives dozens of calls each year reporting street dogfights. Once considered to be an isolated animal welfare issue, dogfighting is now recognized as intricately linked with drug dealing, gambling, gangs, and organized crime. In Birmingham, youth gangs will look to 'settle scores' with impromptu dogfights. In Detroit, Michigan, dog-fighting is a whole sub-economy, as the money received from a dog-fight is easily more than that gained from an armed robbery.

In crime-ridden neighbourhoods, dogfighting is dangerously pervasive. One report found that almost all of the fourteen-year-olds at a public high school in Michigan had personally seen a dog-fight. Chicago's Animal Abuse Control Team reports that eight-year-olds have been seen conducting their own dogfights.

Though it is an illegal activity linked to crime, the people supporting the crime are not necessarily criminals. One former gang member who turned youth worker said,

> Lots of people come from all over, Canada, suburbs, and upper class blacks and brown men. . . . They come to see dog fights to the death. It is . . . like a carnival. . . . What I hate, they blame it all on the ghetto, it's really lots of people not from the ghetto that make dog fighting happen in Detroit.

The range of dogfighters also varies. Among street fighters, fighting dogs are status symbols and weapons for gang members; there are also mid-level fighters in rural areas. High-end professionals hold fights on the national and international level, with strict rules and regulations. They carefully breed generations of fighting dogs whose stud services command enormous prices. They publish news of thousands of fights in journals and over the Internet, and also announce fights to come.

The one thing dogfights have in common is the suffering of the dogs. Their lives are short and violent. Losing dogs who are not killed in the ring are usually killed by their owners, since they must be punished for the embarrassment of losing the fight. Even winning dogs frequently die from their wounds.

And it is not only the fighting dogs who suffer. In Arizona in 2004, Mike Duffey from the Pima County Sheriff's Department began matching the bodies of dead dogs to reports of missing pets. More than 3,000 dogs were reported missing, and Duffey estimated that as many as half of them were stolen from people's homes then used as 'bait', sacrificed to fighting dogs. After they were mauled to death, their bodies were dumped in the desert.

How Attached to You Is Your Dog?

Despite the abuse dogs can suffer at our hands, no other species is as loyal to the human race as the dog. Throughout the centuries, their devotion has not gone unnoticed. The nineteenth-century writer Josh Billings said, 'A dog is the only thing on earth that loves you more than he loves himself.'

On one spring morning in 2000, eight-year-old Steven, his dog Elmo, and his friend Ethan were hunting for frogs in a pond in the middle of the woods in Ontario, Canada. On their way home, they became lost in a swamp full of quicksand and treacherous waters.

Ethan became entangled in vines and branches, and a frightened Steven went to get help. Instead of following his owner, Elmo stayed with Ethan.

Steven finally made his way home, and a search party was sent out for Ethan. They searched all day, but there was no sign of the lost boy or the dog. The search party recruited a helicopter and a firefighting team. Still no sign. As night was falling, the rescue team was about to give up when a firefighter saw a pair of eyes reflected in the light. It was a cold and shivering Elmo. Ethan was almost unconscious with hypothermia, and Elmo had huddled by his side, keeping him warm and saving his life.

Scientists are trying to fathom the extent of a dog's devotion. David Tuber from Ohio State University gave dogs a choice to spend time either with him or with their kennel mates. The dogs had lived with their kennel mates since they were eight weeks old. They played together, ate together, and slept together. Tuber only interacted with the dogs to feed them, clean their runs, and take them out to exercise. Yet, when given a choice of whom to spend time with, the dogs preferred to spend time with Tuber rather than their kennel mates. When the dogs were moved to a new and potentially stressful environment, their glucocorticoid (a stress hormone) levels were lower if Tuber was present than if they were with kennel mates.

Not only do dogs prefer to spend time with a human than with one of their own species, but they are so focused on us that sometimes it can work to their disadvantage. For instance, we know how much dogs love their food, and if they have to choose between a small pile of food and a large pile, they usually choose the larger pile. But if your dog sees you repeatedly choose the small amount of food, they are more likely to choose the small amount.

And if you put meat under one cup but choose the opposite cup, they will choose the wrong cup more often. Even when they can smell which cup the food is under, and can see which cup you put the food under, against their better judgement they will choose the cup you point to, just because they trust you.

These experiments led Tuber and other researchers to conclude that the bond dogs have with humans is similar to the bond children have with their parents. József Topál from Eötvös Loránd University in Hungary decided to put this theory to the test.

After World War II, thousands of orphans were left homeless, hospitalized, or institutionalized throughout Europe. The World Health Organization wanted to measure the social impact of these orphans and commissioned a report on what happens to the psychology of infants when they are deprived of their mothers.

A few decades later, Mary Ainsworth followed up on this research by developing a test called 'Strange Situation' to evaluate the relationship between a mother and her child. It is a kind of TV series with several episodes where the mother and a child between the age of six months and two years arrive at a playroom. A stranger enters and the mother leaves while the stranger plays with the child. Then the mother returns. The child is left completely alone, then the mother and stranger return together.

Ainsworth found that children tend to use their mother as a secure base to explore the playroom and the new toys. They were less likely to explore when their mother left the room or when the stranger was present.

But even more interesting was what happened when the mother returned after a short absence. Most children were 'securely attached', where they were happy to see their mother and greeted her with hugs and kisses. However, some were 'anxious avoidant', ignoring her and acting resentfully. Others had an even more severe reaction that was called 'anxious resistant', where they greeted their mother's return with anger, kicking, squirming, and refusing to be held.

The way in which children interact with their mother or a primary caregiver is described as *attachment theory*. Many researchers have explored attachment in other animals, but Topál wanted to explore attachment between two different species – humans and dogs.

Many people see their dogs as their children and call themselves 'parents' rather than owners. Dogs behave like children in various ways: following their owners around, vocalizing to get their attention, and clinging to them when they are unsure. Topál decided to use the Strange Situation test on dogs.

Researchers kept the study as close to the human model as possible. The dog and owner were introduced to a playroom. They were joined by a stranger; then the owner left while the stranger played with the dog. The owner returned and then left the dog alone. Finally, both the owner and the stranger returned.

Dogs were similar to children in that they explored and played more when their owners were in the room. Just as children showed searching behaviour when their mothers left, dogs stood at the door when their owners left the room. Other researchers found that when owners left, their dogs showed extreme searching behaviours such as barking and scratching at the door.

Upon the owners' return, the dogs were more like the securely attached infants, seeking physical contact almost immediately with contented behaviour like tail wagging. Topál concluded that the attachment of dogs to their owners is similar to the attachment of infants to their mothers.

Marta Gácsi from Eötvös Loránd University then tried the Strange Situation test with rescue dogs. Instead of an owner, they used someone who had only played with the dog for ten minutes a day over three days. Even after such short interactions, the dogs stayed by the familiar person and did not try to escape or follow the stranger out of the room. When the familiar person returned after an absence, the dogs also approached that person more often than the stranger. It seems that the bond between humans and dogs forms quickly, and even short interactions with a stranger can result in attachment.

DogMatchmakers.com

Dogs show an affiliation towards humans that is unlike any other in the animal kingdom. They prefer humans to their own species and can behave like human infants towards their parents. If the dictionary defines *love* as 'a feeling of warm, personal attachment, or deep affection', then this is definitely what dogs have for us.

Happily, their love is far from unrequited. There are 78 million dogs in the US and 11 million dogs in the UK. Numbers vary, but up to 81 percent of people view their pets as family members and think about their dogs as much as their children. In a few cities, including San Francisco, California, there are more dogs than children. Dog owners have formed a US political action committee, and the 'dog vote' is fully expecting to influence the city's upcoming mayoral election. Both men and women show similar empathic responses to stories of babies and puppies in pain. More than half of dog owners call themselves 'Mummy' or 'Daddy', and 71 percent have a photo of their dog in their wallet or phone that they show to other people. The special status of dogs comes with benefits – in the US 62 percent have their own chair, sofa, or bed; 13 percent have their own room; and 55 percent have received at least one birthday present, with others getting cakes and parties. Almost a quarter of dog owners have taken a sick day to take care of their dog.

Not surprisingly, myriad companies provide a range of pet services. In 2010, the American pet industry was worth $48 billion (£30 billion). The British spent about £2.7 billion on their pets. During the economic crisis of 2008, it was one of the only industries that showed growth – a healthy 5 percent.

We love dogs so much that they have become a part of every facet of our lives, including finding romance. In fact, to all of the people out there hoping to find that special someone, I have one piece of advice for you – get a dog.

On a sunny afternoon in July, a young handsome Frenchman stood on a street corner in Vannes, a beautiful spot on the west Atlantic coast in Brittany. Every now and then a woman would walk past, and he would say, 'Hello, my name's Antoine. I just want to say that I think you're really pretty. I have to go to work this afternoon, but I was wondering if you would give me your phone number. I'll ring you later, and we can have a drink together someplace.'

Antoine gazed meaningfully into her eyes and offered a winning smile. He did this throughout the summer and was sometimes accompanied by a black, friendly-looking mutt.

The difference between Antoine and other young Frenchmen fishing for phone numbers was that if the woman agreed and gave him her number, instead of taking her out, Antoine revealed that she had been part of a study about courtship behaviour. (Except for one lucky girl whom Antoine *did* call. They ended up getting married.)

The best predictor of Antoine's success? The black mutt standing innocently by his side. If Antoine was alone, despite his good looks, only 9 percent of women gave in to his smile. If the dog was present, Antoine's rate of success went up to a whopping 28 percent – almost one in three. Most men would happily take those odds.

But perhaps Antoine's success was because his dog was particularly good at getting the ladies' attention. Maybe the dog tugged on her skirt or had an irresistibly cute collar.

In another study (described in Chapter 9 earlier), a young American university student used a black Labrador who was a guide dog and therefore trained not to interact with people or solicit attention in any way. In one condition, the dog was cute in a matching lead and collar, and in another, the dog was intimidating in a black-studded leather collar and a frayed rope for a lead. The owner also had an outfit change. In one condition, he looked smart in collared shirt and tie with a sports jacket. In the other, he wore an old T-shirt, torn, dirty jeans, and scuffed work boots.

The best predictor of success? Again, whether the young man had the dog. It did not matter if the dog wore a studded collar or not, or whether the guy looked sharp or like a bum. If the dog was there, the number of people who smiled or talked to him went up by 1,000 percent.

Of course, some dogs are more inviting than others, and having an eight-week-old golden retriever at the end of your lead is more likely to get you smiles, approaches, and phone numbers than a giant, slobbering Rottweiler (no offence to Rottweiler owners, but someone did the study, and the golden retriever puppy was by far the winner). But in the end, a dog is pretty much a dog, and no matter if you are scruffily dressed or smart, a handsome Frenchman or a casual American, a dog will have all sorts of strangers coming up to say hello. As a tip to those posting profile photos on Match.com – even being photographed with a dog can make you look happier, more relaxed, and more approachable.

A Cure for Lonely Hearts

Dogs can be rewarding in all sorts of situations besides looking for love. Almost 90 percent of people report that their relationships with other people make their life meaningful. Yet we are becoming increasingly isolated, and much of our contact with other people is made through a computer. Loneliness is something we all experience from time to time. Intense loneliness can be painful, frightening, and lead to low self-esteem and feelings of hopelessness. Loneliness can put you at risk for headaches, ulcers, and sleep deprivation, which in turn can lead to car accidents, drinking problems, and even suicide. One author even found that loneliness added to your risk of a heart attack.

Over the years there has been a ton of research that says that having a pet, a dog in particular, can make people less lonely, particularly among the elderly, single women, children, and gay men. Children in wheelchairs are much more likely to receive smiles and friendly glances, and even have other children start a conversation, if they have a service dog with them.

But when some researchers went so far as to recommend pets to alleviate loneliness, others ran studies that challenged their results. Andrew Gilbey from Massey University in New Zealand and colleagues assessed people on a questionnaire designed to test how lonely they were. After six months, more than half the people had adopted a pet (almost half of these pets were dogs), and Gilbey reassessed them to find out if they really were less lonely. There was no overall change in loneliness in people whether or not they had acquired a pet, and whether this pet was a cat or a dog. In fact, in the group that did get a pet, there was a slight increase in loneliness.

There is no doubt, however, that people *believe* their pets make them less lonely. Many of us turn to our dogs when we are in need of comfort or support. A study of 401 university students showed that in times of emotional distress, they were more likely to turn to their dog than to their fathers or brothers for support. Another study found that after experiencing rejection, thinking about a pet was just as effective as thinking about a best friend in staving off negative feelings. Yet another study found that a group of women actually performed better at a cognitive task in the company of their dog rather than their best friend. Finally, when comparing pets with romantic relationships, people's relationships with their pets were more secure. For instance, when read the statement 'I know my pet really loves me', 52 percent of people agreed; the statement 'I know my partner really loves me' brought only 39 percent agreement.

In New York, forty-eight stressed-out stockbrokers were taking medication for high blood pressure. The drug lowered their blood pressure, but only when they were resting – their blood pressure still shot up when they were confronted with a stressful situation. When some of these stockbrokers adopted a pet, either a cat or a dog, they found their blood pressure was lower, even when a stressful situation presented itself. They even performed better at mathematical problems with their pets present.

The Healing Power of Dogs

When it comes to more serious health issues, there is a lively debate going on as to whether owning a pet helps or hinders. Throughout history, dogs have been used as healers. In the town of Epidaurus on the Saronic Gulf, there is a temple built to Asklepios, the son of Apollo. The temple was an ancient Greek version of a health spa. After a long day of purification and sacrifice, patients would go to sleep deep within the temple chambers. During the night, the god would supposedly come in the form of a dog and lick their wounds. Stone tablets record a list of people reportedly healed this way, including a blind boy whose sight was restored when he was licked across the eyes and another who was cured of a large growth on his neck.

During the reign of Queen Elizabeth I, doctors recommended lapdogs as a cure for various ladies' illnesses. For example, holding a dog to the bosom was supposedly a good cure for a weak stomach.

With the advent of modern medicine, science turned up its nose at using animals as treatment. For more than a hundred years, animals were seen as unclean. The only time dogs were referred to was to emphasize the danger of transferable diseases and the menace they presented to public health.

This all changed in 1980 with a single study by Erika Friedmann from Brooklyn College in New York. Friedmann and her colleagues reported that after people were hospitalized for a heart attack, those with pets were 23 percent more likely than those without pets to still be alive a year later. This was the first study published in a medical journal to show that owning an animal contributed to the prevention of disease. The researchers replicated their study fifteen years later, and specifically examined dogs rather than pets in general, and found that dog owners were significantly less likely to die in the year after a heart attack than non-dog-owners. (Cat owners, by the way, were no more likely to survive than non-owners.)

In 1992, a study in Australia found that pet owners had lower blood pressure, cholesterol, and triglycerides (the main ingredient of animal fat) than non-pet-owners. Even more recently, in 2001, people with pets not only had lower resting heart rates and blood pressure, but when they were presented with a stressful arithmetic problem or had their hand plunged in iced water, they experienced smaller increases in heart rate and blood pressure, as well as a faster recovery.

For the elderly, stressful life events like a loved one dying usually translate into more frequent visits to the doctor. Elderly people who owned pets reported fewer doctor visits than non-owners, even in difficult times.

It is no wonder that in 1987, the US National Institutes of Health released the following recommendation:

> All future studies of human health should consider the presence or absence of a pet in the home. . . . No future study of human health should be considered comprehensive if the animals with which they share their lives are not included.

This recommendation prompted a backlash, with several researchers publishing studies that said owning a pet did nothing at all for your health. The Australian study that found people with pets had lower blood pressure was refuted by another study, which argued that people who turned up to free cardiovascular screenings tend to have less body fat, enjoy lower blood pressure, and smoke and drink less than those who avoid those screenings. To prevent this self-selection in the data, researchers randomly called households and asked what kind of pet they had and whether they had any habits that put them at high risk for heart disease. They found that pet owners actually had higher blood pressure, more body fat, and were more likely to be smokers.

Another study found that dog owners had no advantage over non-dog-owners when it came to avoiding a heart attack, and cat owners were *more* likely to have a heart-related death or be readmitted into hospital. They went so far as to call Friedmann's original study 'a canard' (French for 'duck', used to refer to a report as false).

There is one context, however, where dogs seem to have

indisputable benefits to human health. In Animal-Assisted Therapy, an animal becomes part of a treatment plan.

David Beck* was a forty-three-year-old man who lived alone on disability for bipolar disorder. He was single and could not hold down a job. He lost his mother at an early age, and the only fond memories he had from his childhood were of his pet dog.

One day he was assaulted by a group of thugs who punched him, slammed his head on the ground, and stole his most precious possession, his guitar. In the days that followed, David could not stop reliving the events in his head. He fell into a deep depression, could barely talk, could not sleep, and was constantly tearful. He developed manic symptoms, which got him in trouble with the police. Finally, he ended up checking into the psychiatric ward of a hospital. Doctors prescribed him various mood-stabilizing medications, but nothing seemed to work. David was unable to perform even the most basic tasks to take care of himself.

Help came to David in the form of a golden retriever called Ruby. She spent several hours a day with David for three weeks. During that time, doctors told David that Ruby was his responsibility. He had to walk her, groom her, and take care of her whenever they were together.

The result was a remarkable recovery. David's moods began to improve. He began to talk again. He felt less anxious, slept through the night, and stopped the repetitive jerking movements that had been so unsettling. He even attracted the attention of a few women, who approached him to talk about Ruby.

Not only did Ruby help David, she also improved the mood on

---

* This is not the patient's real name.

the cardiovascular, neurological, and surgery floors where she visited, among both staff and patients. Two patients who were not responding to medication became more motivated and less withdrawn after visits from Ruby.

After three weeks of treatment, David's doctors decided even more time with Ruby would be beneficial. With Ruby by his side, David managed to find a flat, contact old friends, and take care of himself. For the next year, even after he checked out of hospital, David would spend time with Ruby, which he always enjoyed.

Animal-Assisted Therapy is rapidly gaining support among health care practitioners. There are not many empirical studies, but the literature is growing, and most of it seems to indicate that the therapy has mostly positive results, especially with children.

Hospitals can be traumatic for the young, particularly when they are separated from their parents and must undergo painful procedures. A group of children who were hospitalized for a range of reasons, including cystic fibrosis, transplants, and cancer, interacted with a therapy dog one night a week. Another group had a ninety-minute play session with toys and other children. Seeing the therapy dog made children more excited than a roomful of toys and other children. Parents thought their children seemed happier after the therapy dog visit than the play session.

Children in the pediatric ward of St Cloud Hospital spent either fifteen minutes resting, or fifteen minutes working with a therapy dog. Those who worked with the therapy dog reported that their pain was four times less than children who just relaxed quietly – comparable to a dose of paracetamol. One child who worked with a therapy dog had a reduction in pain level from eight to zero, without any drugs, for at least three hours.

*

Even for something as routine as a doctor's examination, children showed much less distress, such as crying and screaming, and were less likely to have to be restrained when a dog was present.

Animal Assisted Therapy works with grown-ups, too. People with dementia who experienced a worsening of symptoms after sunset (sundowner's syndrome) were less agitated and aggressive when they interacted with a therapy dog. A single visit from a therapy dog was more effective than a recreation session in reducing anxiety in psychiatric patients. Cancer patients rated visits from a therapy dog as being just as comforting as a human visit and more likely to make them feel better.

The list goes on. While originally scorned by the medical profession as a health threat, now dogs are proving to be effective as a treatment for a vast range of conditions and are occasionally more effective than drugs.

The Biology of Love

What about those of us who are not stressed or lonely, and do not have heart disease or a terminal illness? Like the caricature of the old cat lady, our obsession with dogs has spawned its own caricature. 'Dog people' speak to their dogs like children, dress them in ridiculous clothes, and leave them large sums of money in their estates. Their dogs are their best friends because no one else will volunteer.

However, the profile of an ordinary dog owner (or cat owner!) is not some sad, lonely person looking to make a human connection and coming up short. In contrast, people who own pets tend to be more extroverted, less lonely, and have higher self-esteem than people who do not own pets. Far from seeing their pets replacing their key relationships, pet owners are just as close to best friends, parents, and siblings as non-owners. Pets are an extra form of social support, not a replacement. And even if we do not dress our dogs in Ralph Lauren, include them in the family portrait, or talk to them when we think no one is listening, our relationship with dogs is so deeply entrenched that it changes our very physiology.

One hormone that seems to be particularly affected in our relationship with our dogs is oxytocin. Oxytocin is transmitted from the brain directly into the bloodstream and along nerve fibres to the nervous system. Sometimes called the 'hug hormone', oxytocin is what makes you feel good when you are touched by a loved one, get a massage, or enjoy a good meal.

In a study from Japan, people whose dogs gazed at them for a longer period of time showed a higher increase in oxytocin than people whose dogs gazed at them for a short amount of time. Not only that, but people with dogs who looked at them longer reported being happier with their dogs than those whose dogs did not look at them for as long.

In another study, people were brought into a room that was empty except for two tables and two chairs. They sat on a rug on the floor with their dogs and had their blood drawn by a nurse. For the next thirty minutes, the owners' attention was completely focused on their dog. They talked softly to their dog, stroked her gently, and scratched her body and behind her ears. Their blood was drawn again after thirty minutes.

People's blood pressure decreased and they experienced an increase in oxytocin as well as in a whole range of hormones, including beta-endorphins, which are associated with euphoria and pain relief; prolactin, which promotes bonding and is associated with parenting behaviour; phenylethylamine, which tends to increase when people find a romantic partner; and dopamine, which increases pleasurable sensations.

When the humans came in and read a book for thirty minutes, oxytocin and the other hormones did not increase as much as was seen following interaction with their dogs. What is even more incredible is that not only did humans experience a rise in these hormones – the dogs did, too! It seems the feelings of bonding and affiliation are entirely mutual. No one has done the same studies between humans and other animals, but I predict that there would not be the same changes in all of these hormones with any other species.

Suzanne Miller and colleagues from Colorado State University wanted to see whether there was a difference in how women and men were affected by dogs. Again, people were asked to either interact with their dog or read for twenty-five minutes. Miller found that women who played with their dog had a 58 percent increase in oxytocin. Men, however, had a 21 percent decrease in oxytocin. This does not mean women get more enjoyment out of petting their dogs than do men. Though oxytocin is not strictly a female hormone, it is linked to the female sex hormone oestrogen, and its effects on women are more pronounced. Also, although men's oxytocin decreased while petting their dog, in the reading condition it decreased by twice as much.

Hormones in men that might change as a result of interacting with their dog probably include hormones such as testosterone. Testosterone is found at levels ten times greater in men than women and is related to changes that occur during puberty such as developing muscle and body hair. It is also associated with striving for status. Men's testosterone typically goes up when they are getting ready to compete.

Before a canine agility competition, researchers analysed the testosterone and cortisol in the saliva of eighty-three men and their dogs. The teams performed in the competition, and after the results were announced, the men and dogs gave saliva again to be analysed.

Men who had higher levels of testosterone before the competition were more likely to win. But if men with high levels of testosterone happened to lose, they became more frustrated than men with lower levels of testosterone who lost. In the agility competition, once you lost, you were unable to compete again. Thus status-striving men with high testosterone not only lost status by losing the competition, but were unable to regain their status by competing again.

What was surprising is that after losing the competition, the dogs of these high-testosterone males had higher levels of cortisol, a stress hormone. The researchers looked into why human testosterone would have an effect on a dog's cortisol. It turns out that men who had lower testosterone before the competition were more likely to pet their dog and play with them afterwards. However, high-testosterone males showed more aggressive behaviour towards their dogs after their loss and were more likely to push their dog and shout at them. This is probably what led to an increase in their dog's cortisol.

Since the physiologies of men and women are affected by dogs in different ways, are dogs affected in different ways by men and women? Many people believe that their dogs react differently to humans depending on their gender. And at least in one study, dogs tended to show more aggressive reactions towards men, such as barking and maintaining eye contact.

In an animal rescue home in Dayton, Ohio, dogs were taken into a room in the centre. There someone gently restrained them and extended one of their front legs so a vet could draw blood with a syringe. The people petted the dogs for twenty minutes, then extended the leg for a second blood draw.

Just being petted had a positive effect on the dogs (something to remember when your dog is at the vet). But dogs petted by females showed a decrease in cortisol between the two blood draws, while dogs petted by males showed an increase in cortisol. What was it about females that made the dogs less stressed? Was it something about the way women smell or the way they look? Or is it something about the way women tend to interact with dogs?

In a following study, the men underwent 'petting training' to ensure they petted the dogs the same way as the women. They were instructed to massage the dog's shoulder, back, and neck muscles deeply and to give long, firm strokes from the dog's head to their back legs. Throughout the massage, men were to talk to the dog in a calm and soothing voice.

If dogs responded better to women because of the way women interacted with dogs, then training men to act in the same way should have the same effect. If it was something hormonal or more subtle, then it would not matter how men behaved; the dogs would still be less stressed with women as opposed to men.

After training men to pet dogs more like women, the men could reduce the cortisol in the dogs as effectively as women. Showing that, as with women, dogs respond better to men if they try a little tenderness.

Dogs have such a natural affinity to humans that the gentle stroking of a human hand can release chemicals inside their brains that make them feel calm and affectionate. They even prefer to be with humans than with their own species. In return for a lifetime of loyalty, they depend on us for food, the warmth of a loving family, and a good home. It is up to us to uphold our end of the bargain. Dogs deserve it – they are geniuses, after all!

ACKNOWLEDGEMENTS

We could not have written this book without Vanessa's mom, 'Bobo' (Jacquie Leong), who turned up from Australia the day our baby was born and stayed for eight months to help while we wrote the book (then stayed up all night copyediting it!). Loving thanks to Brian's folks, 'Mema' and 'Pops' (Alice and Bill Hare), who started it all by bringing home a wriggling black Labrador called Oreo. Also thanks to both of our mothers for their meticulous proofreading of the entire text.

Brian would never have considered writing this book if it were not for the encouragement of his good friend and colleague Terence Burnham (who was the first to suggest the idea of the book and the title while Brian was still in graduate school). Thanks to Richard Wrangham and Mike Tomasello for reading an earlier draft of several chapters and being the best mentors a young scientist could hope for. A special thanks to Irene Plyusnina for hosting Brian in Novosibirsk. She and Viktor treated Brian like he was part of the family. Thank you to Brian's students, who have shaped much of our thinking about dog cognition through all our discussions together at the Duke Canine Cognition Center (www.dukedogs.com). In particular, thanks to Brian's colleagues and graduate students Victoria Wobber, Alexandra Rosati, Evan Maclean, Jingzhi Tan, Kara Schroepfer, Courtney Rainey, Chris Krupenye, and Korrina Duffy, from whom we have learned so much and who patiently tolerated Brian's split attention while he was working on this book. Also thanks to Emily Bray, Zoey Best, Mary Dambro, Isabel Bernstein, and Ashton Madison, who helped us track down and organize all of the references we cite during the summer of 2011. Thank you to the National Science Foundation (NSF-BCS-1025172), the Eunice Kennedy Shriver National Institute of Child Health and Human Development, the Mars' Waltham Centre (R03HD070649), and the Office of Naval Research (N00014-12-1-0095) whose support in part made this book possible. Thank you to our book agent, Max Brockman, whose advice was invaluable as we initially began brainstorming about what our book might be about. Thank you to our amazing editor Stephen Morrow, who took a chance on a husband–wife team and skilfully guided us through the editorial process. A much-improved book resulted from his efforts. Thanks also to Stephanie Hitchcock and LeeAnn Pemberton at Dutton for their meticulous work throughout, in particular on our endnotes. Finally, thanks to Bryan Golden (kagomesarrow87@yahoo.com) for the beautiful drawings included in the text.

We hope our book will inspire all humans to show even more compassion towards the animals with whom we share this Earth. There are millions of dogs who are much less fortunate than your beloved pup. Anyone wishing to help the millions of dogs and cats who end up in shelters every year should adopt their next pet and in the meantime donate to the Humane Society International: www.hsi.org. If you wish to help Claudine André with her mission to save bonobos and to encourage the youth of Congo to show kindness to humans and all animals – including dogs – please consider donating to Friends of Bonobos: www.friendsofbonobos.org. If you wish to support research projects aimed at helping animals, consider donating to a research project of your choosing at www.petridish.org. If you would like to follow our research group's progress, you can visit www.dukedogs.com. You will find a page with links to all of the other researchers around the world who are studying dog cognition.

Finally, to find out how you can use the new science of dognition to discover your own dog's unique genius, visit www.dognition.co.uk.

NOTES

Preface (pages ix–xiii)

ix–x   **Rico who had solved a similar problem** Kaminski, J., J. Call, and J. Fischer,
        "Word Learning in a Domestic Dog: Evidence for 'Fast Mapping,'" *Science*
        304, no. 5677 (2004): 1682–83.

    x   **Language-trained bonobos, bottlenose dolphins, and African grey
         parrots** Savage-Rumbaugh, E. S., et al., "Language Comprehension in Ape
         and Child," *Monographs of the Society for Research in Child Development* 58,
         nos. 3–4 (1993).

         and, Herman, L. M., "Exploring the Cognitive World of the Bottlenosed
         Dolphin." In *The Cognitive Animal: Empirical and Theoretical Perspectives
         on Animal Cognition*, eds. M. Bekoff, C. Allen, and G. M. Burghardt
         (Cambridge, Mass.: MIT Press, 2002).

         and, Pepperberg, I. M., *The Alex Studies: Cognitive and Communicative
         Abilities of Grey Parrots* (Cambridge, Mass.: Harvard University Press, 2002).

Chapter 1: Genius in Dogs? (pages 3–15)

    5   **Stanford-Binet Intelligence Scale, which is known as the IQ test** Hunt,
         M. M., *The Story of Psychology* (New York: Anchor Books, 2007).

    5   **dropped out of university and became billionaires** "List of college dropout
         billionaires," *Wikipedia*, last modified October 5, 2012, http://en.wikipedia
         .org/wiki/List_of_college_dropout_billionaires.

    5   **'No, not exceptionally. Instead he was a genius'.** Issacson, W., *Steve Jobs*
         (New York: Simon & Schuster, 2011).

    5   **reduce the stress of millions of farm animals** Gregory, N. G., and
         T. Grandin, *Animal Welfare and Meat Science* (CABI Publishing, 1998).

and, Grandin, T. and C. Johnson, *Animals Make Us Human: Creating the Best Life for Animals* (New York: Houghton-Mifflin Harcourt, 2009).

6   **revolutions seemed to have happened, the 1960s** see Hunt, *The Story of Psychology.*

6   **cognitive abilities that are not necessarily interdependent on one another** see Ibid.

and, Bekoff, *The Cognitive Animal.*

and, Hauser, M. D., *Wild Minds: What Animals Really Think* (New York: Owl Books, 2001).

6   **other types of memory are equally as good** Squire, L. R., "Memory Systems of the Brain: A Brief History and Current Perspective," *Neurobiology of Learning and Memory* 82, no. 3 (2004): 171–77.

and, Roediger, H. L., E. J. Marsh, and S. C. Lee, "Varieties of Memory." In *Stevens' Handbook of Experimental Psychology,* 3rd ed. vol. 2, *Memory and Cognitive Processes* (2002): 1–41.

7   **uncanny precision – even down to the time of day** Parker, E. S., L. Cahill, and J. L. McGaugh, "A Case of Unusual Autobiographical Remembering," *Neurocase* 12, no. 1 (2006): 35–49.

7   **spatial problems requiring navigation between landmarks** Maguire, E. A., K. Woollett, and H. J. Spiers, "London Taxi Drivers and Bus Drivers: A Structural MRI and Neuropsychological Analysis," *Hippocampus* 16, no. 12 (2006): 1091–101.

8   **the same distance it takes to get to the moon** Egevang, C., et al., "Tracking of Arctic Terns *Sterna paradisaea* Reveals Longest Animal Migration," *Proceedings of the National Academy of Sciences* 107, no. 5 (2010): 2078–81.

8   **heartier dinner than if they hunted alone** Wiley, D., et al., "Underwater Components of Humpback Whale Bubble-net Feeding Behaviour," *Behaviour* 148, nos. 5–6 (2011): 575–602.

8   **bees where to find nectar-filled flowers** Esch, H., "Foraging Honey Bees: How Foragers Determine and Transmit Information About Feeding Site Locations." In *Honeybee Neurobiology and Behavior,* eds. C. Giovanni Galizia, Dorothea Eisenhardt, Martin Giurfa (New York: Springer, 2012), 53–64.

8   **we can identify different types of animal genius** Kamil, A. C., "On the Proper Definition of Cognitive Ethology." In *Animal Cognition in Nature,* eds. R. P. Balda, I. M. Pepperberg, and A. C. Kamil (San Diego: Academic Press, 1998), 1–28.

and, MacLean, E. L., et al., "How Does Cognition Evolve? Phylogenetic Comparative Psychology," *Animal Cognition* 15, no. 2 (2012): 223–38.

9   **nine months before, even though the seeds are covered in snow** Tomback, D. F., "How Nutcrackers Find Their Seed Stores," *The Condor* 82, no. 1 (1980): 10–19.

9   **the champions of finding food they had hidden** Kamil, A. C., and R. P. Balda, "Cache Recovery and Spatial Memory in Clark's Nutcrackers (*Nucifraga columbiana*)," *Journal of Experimental Psychology: Animal Behavior Processes* 11, no. 1 (1985): 95–111.

and, Balda, R. P., A. Kamil, and P. A. Bednekoff, "Predicting Cognitive Capacity from Natural History: Examples from Four Species of Corvids," *Papers in Behavior and Biological Sciences* 13 (1996): 15.

and, Bednekoff, P. A., et al., "Long-Term Spatial Memory in Four Seed-Caching Corvid Species," *Animal Behaviour* 53, no. 2 (1997): 335–41.

9 **remember where other birds had hidden food, scrub jays proved themselves masters** De Kort, S. R., and N. S. Clayton, "An Evolutionary Perspective on Caching by Corvids," *Proceedings of the Royal Society B: Biological Sciences* 273, no. 1585 (2006): 417–23.

9 **nutcrackers were hopeless in the same situation** Bednekoff, P. A., and R. P. Balda, "Observational Spatial Memory in Clark's Nutcrackers and Mexican Jays," *Animal Behaviour* 52, no. 4 (1996): 833–39.

9 **food in darker locations to avoid allowing others to see them cache it** Dally, J. M., N. J. Emery, and N. S. Clayton, "Food-Caching Western Scrub-Jays Keep Track of Who Was Watching When," *Science* 312, no. 5780 (2006): 1662–65.

and, Grodzinski, U., and N. S. Clayton, "Problems Faced by Food-Caching Corvids and the Evolution of Cognitive Solutions," *Philosophical Transactions of the Royal Society B: Biological Sciences* 365, no. 1542 (2010): 977–87.

9–10 **allowing others to cross over on their backs** Anderson, C., G. Theraulaz, and J. L. Deneubourg, "Self-Assemblages in Insect Societies," *Insectes Sociaux* 49, no. 2 (2002): 99–110.

10 **One mistake can mean death** Cheney, D. L., and R. M. Seyfarth, *Baboon Metaphysics: The Evolution of a Social Mind* (Chicago: University of Chicago Press, 2007).

11 **new version of a problem they have seen before** Tomasello, M., and J. Call, *Primate Cognition* (New York: Oxford University Press, 1997).

11 **no water was visible when she thought of her solution** Tennie, C., J. Call, and M. Tomasello, "Evidence for Emulation in Chimpanzees in Social Settings Using the Floating Peanut Task," *PLoS ONE* 5, no. 5 (2010): e10544.

and, Hanus, D., et al., "Comparing the Performances of Apes (*Gorilla gorilla, Pan troglodytes, Pongo pygmaeus*) and Human Children (*Homo sapiens*) in the Floating Peanut Task," *PLoS ONE* 6, no. 6 (2011): e19555.

12 **that were her toys and those that were non-toys** Pilley, J. W., and A. K. Reid, "Border Collie Comprehends Object Names as Verbal Referents," *Behavioural Processes* 86 (2011): 184–95.

12 **some dogs spontaneously inferred the toy must be in the other cup** Erdőhegyi, Á., et al., "Dog-Logic: Inferential Reasoning in a Two-way Choice Task and Its Restricted Use," *Animal Behaviour* 74, no. 4 (2007): 725–37.

and, Aust, U., et al., "Inferential Reasoning by Exclusion in Pigeons, Dogs, and Humans," *Animal Cognition* 11, no. 4 (2008): 587–97.

13 **Warner reported on a German shepherd named Fellow** Warden, C., and L. H. Warner, "The Sensory Capacities and Intelligence of Dogs, with a Report on the Ability of the Noted Dog 'Fellow' to Respond to Verbal Stimuli," *The Quarterly Review of Biology* 3, no. 1 (1928): 1–28.

14   **In the intervening seventy-five years, dogs were largely ignored** At the end of
      the nineteenth century, dogs were a central character in studies of animal
      behaviour and intelligence. They were extremely familiar to scientists, since
      dog breeding was then at the height of popularity in Europe. Charles Darwin
      himself was an avid hunter and dog lover. Upon his return from his world tour
      on the HMS *Beagle* in 1836, he was overjoyed to be reunited with his dogs at
      Down House. He spent the remainder of his years in the almost constant
      company of dogs (Townshend, E., *Darwin's Dogs: How Darwin's Pets Helped
      Form a World-Changing Theory of Evolution* [London: Frances Lincoln Ltd.,
      2009]). He spent a great deal of time observing and watching his dogs as he
      formulated his ideas about variation in animals (see Darwin, C., *The Variation
      of Animals and Plants under Domestication,* vol. 1 [London: John Murray,
      1868]) as well as animal communication and emotions (Darwin, C., *The
      Expression of the Emotions in Man and Animals* [John Murray, 1872]). His
      colleague George Romanes wrote two volumes exploring the potential of
      animal intelligence. In these volumes he relayed a series of stories that he had
      gathered from a variety of dog experts. While he clearly noted the importance
      of using experiments to test ideas about animal intelligence, his collection of
      animal stories attributed human-like abilities to a variety of animals (Romanes,
      G. J., *Mental Evolution in Animals: With a Posthumous Essay on Instinct by
      Charles Darwin,* ed. C. Darwin [London: Kegan Paul, Trench, 1883]; and,
      Romanes, G. J., *Animal Intelligence,* vol. 44 [London: Appleton, 1883]).
          Shortly after, as psychology launched itself as a science, dogs largely
      disappeared from studies of animal intelligence. Early-twentieth-century
      American psychologists believed they had stumbled on a simple, brutal
      truth – there was no mind. Or if there was one, it was irrelevant. Our
      thoughts, emotions, and awareness were mere responses to stimuli (see
      Hunt, *The Story of Psychology*). Extremely sceptical of Romanes's animal
      'just-so stories' and the 'esoteric jibberish' that made up Freud's labyrinthine
      subconscious, there was a cry for something more substantial, more
      measurable, more *scientific.*
          Instead of a mind, there was only what scientists began to call the 'black
      box'. The circuitry inside the box was not important – all you needed to
      know was that if you pushed a button, the box would drive a certain
      behaviour. This new science was called behaviourism, and it just about took
      over psychology in the US until the 1960s (see Watson, J., "Psychology as the
      Behaviorist Views It," *Psychological Review* 20 [1913]: 158–77.) Ironically, it
      was Russian scientist Ivan Pavlov's research on dogs that helped launch
      behaviourism and squelched most research interest in dog intelligence.
      Pavlov noticed that the dogs he was studying had an annoying habit of
      salivating at the sight of their keeper, before there was any food in sight (see
      Hunt, *The Story of Psychology*). One explanation was that dogs realized
      mealtimes were approaching and were salivating in anticipation of the food.
      This hypothesis assumes that dogs are cognitive. But there was another
      explanation – the sight of the person who brought the food was a stimulus
      that caused an automatic physiological response the dogs had no control

over. Pavlov showed that he could induce a dog to salivate in response to any number of stimuli (buzzer, bell, flashing light). When American psychologists learned of Pavlov's dogs, many concluded that good science ruled out the need for richer psychological explanations of behaviour.

The most famous behaviourist of all time, B.F. Skinner, took behaviourism one step further by championing the idea that all behaviour in all animals is explained by a few universal learning principles. He wrote, 'Pigeon, rat, monkey, which is which? It doesn't matter' (Skinner, B. F., "A Case History in Scientific Method," *American Psychologist* 11, no. 5 [1956]: 221–33). While behaviourism banished the need for an animal mind, Skinner trivialized research on species other than rats and pigeons.

This began to change as the cognitive revolution took hold in the 1960s and 1970s. Skinner's universal learning rules floundered when he tried to explain human language. Neuroscientists and computer scientists realized that a cognitive approach was far more powerful in understanding how the brain worked and how to build computers. Scientists studying animals in the wild saw too many signs of intelligence to ignore the possibility that in some ways, they have minds similar to our own (see Bekoff, *The Cognitive Animal*).

But dogs were left out of the equation. Scientists interested in animal cognition were most interested in our close primate relatives. Jane Goodall and Toshisada Nishida found the first evidence of tool use, tool construction, hunting, and lethal territorial aggression in wild chimpanzees (Goodall, J., *The Chimpanzees of Gombe: Patterns of Behavior* [Boston: Harvard University Press, 1986]; and, Nishida, T., *Chimpanzees of the Lakeshore: Natural History and Culture at Mahale* [Cambridge, UK: Cambridge University Press, 2011]). Robert Seyfarth and Dorothy Cheney discovered that the alarm calls of vervet monkeys actually refer to specific predators (Cheney, D. L., and R. M. Seyfarth, *How Monkeys See the World: Inside the Mind of Another Species* [Chicago: University of Chicago Press, 1992]). Frans de Waal observed the deception and forgiveness that makes up chimpanzee political life (De Waal, F., *Chimpanzee Politics: Power and Sex Among Apes* [Baltimore: Johns Hopkins University Press, 2007]). Tetsuro Matsuzawa and Sally Boysen probed the mathematical abilities of chimpanzees (Boysen, S. T., and G. G. Berntson, "Numerical Competence in a Chimpanzee [Pan troglodytes]," *Journal of Comparative Psychology* 103, no. 1 [1989]: 23–31; and, Kawai, N., and T. Matsuzawa, "Cognition: Numerical Memory Span in a Chimpanzee," *Nature* 403, no. 6765 [2000]: 39–40). Sue Savage-Rumbaugh taught a symbolic language to Kanzi the bonobo (see Savage-Rumbaugh, "Language Comprehension") and Mike Tomasello, Josep Call, and Andrew Whiten made observations and conducted experiments to test whether chimpanzees can transmit cultural innovations (Whiten, A., et al., "Cultures in Chimpanzees," *Nature* 399, no. 6737 [1999]: 682–85; and, Whiten, A., V. Horner, and F. B. M. de Waal, "Conformity to Cultural Norms of Tool Use in Chimpanzees," *Nature* 437, no. 7059 [2005]: 737–40; and, Call, J. E., and M. E. Tomasello, *The Gestural Communication of Apes and Monkeys* [Psychology Press, 2007]; and, Tennie, C., J. Call, and M. Tomasello,

"Ratcheting Up the Ratchet: On the Evolution of Cumulative Culture," *Philosophical Transactions of the Royal Society B: Biological Sciences* 364, no. 1528 [2009]: 2405–15). There was an explosion of interest in how monkeys see the world and whether apes live in the shadow of humans. Eventually this enthusiasm for primates did spill over into cognitive research with other charismatic megafauna such as dolphins, parrots, and corvids, but dogs were one of the last species to be swept up in the enthusiasm for studying animal cognition (see Bekoff, *The Cognitive Animal*; and, Shettleworth, S. J., *Cognition, Evolution, and Behavior* [Oxford and New York: Oxford University Press, 2009]; and, Míklósi, Á., *Dog Behaviour, Evolution, and Cognition* [New York: Oxford University Press, 2007], 274).

14 **my dog in my parents' garage and started something new** Hare, B., J. Call, and M. Tomasello, "Communication of Food Location Between Human and Dog (*Canis familiaris*)," *Evolution of Communication* 2, no. 1 (1998): 137–59.

14 **a similar study and independently came to the same conclusion** Miklósi, Á., et al., "Use of Experimenter-Given Cues in Dogs," *Animal Cognition* 1, no. 2 (1998): 113–21.

14 **These studies caused an explosion in the field of dog cognition** see Míklósi, *Dog Behaviour, Evolution, and Cognition.*

Chapter 2: The Wolf Event (pages 17–31)

17 **dogs began evolving from wolves somewhere between 12,000 and 40,000 years ago** vonHoldt, B., et al., "Genome-wide SNP and Haplotype Analyses Reveal a Rich History Underlying Dog Domestication," *Nature* 464, no. 7290 (2010): 898–902.

and, Wayne, R. K., and E. A. Ostrander, "Lessons Learned from the Dog Genome," *Trends in Genetics* 23, no. 11 (2007): 557–67.

and, Morey, D., *Dogs: Domestication and the Development of a Social Bond* (Cambridge, UK: Cambridge University Press, 2010).

17 **which then became domestic dogs over time** Mech, L. D., *The Wolf: The Ecology and Behavior of an Endangered Species*, ed. American Museum of Natural History. Originally published for the American Museum of Natural History (Garden City, N.Y.: Natural History Press, 1970).

and, Mech, D., "Canis Lupus," *Mammalian Species* 37 (1974): 1–6.

and, Clutton-Brock, J., "Man-Made Dogs," *Science* 197, no. 4311 (1977): 1340–42.

and, Serpell, J. A., and E. Paul, "Pets in the Family: An Evolutionary Perspective." In *The Oxford Handbook of Evolutionary Family Psychology*, eds. C. Salmon and T. Shackelford (New York: Oxford University Press, 2011).

18 **since the outcast was no longer thought to be human** Wallner, A., "The Role of Fox, Lynx and Wolf in Mythology," *KORA Bericht* 3 (1998).

18 **Solon of Athens offered a bounty for every wolf killed** Boitani, L. "Wolf Conservation and Recovery." In *Wolves: Behavior, Ecology, and Conservation*, eds. L. D. Mech and L. Boitani (Chicago: University of Chicago Press, 2003), 317–40.

18   The grey wolf's status had been upgraded by 2004 to a species of 'least concern' *The IUCN Red List of Threatened Species* – Canis lupus. 2012 (cited 2012; available from www.iucnredlist.org/apps/redlist/details/3746/0).

18   advised the Japanese to poison all the wolves to protect their livestock Fritts, S. H., R. O. Stephenson, R. D. Hayes, and L. Boitani, "Wolves and Humans." In *Wolves: Behavior, Ecology, and Conservation*, 289–316.

18   This was the last wolf in Japan Walker, B. L., *The Lost Wolves of Japan* (Seattle: University of Washington Press, 2005).

19   them from being hunted and trapped for fur Fritts, S. H., R. O. Stephenson, R. D. Hayes, and L. Boitani, "Wolves and Humans." In *Wolves: Behavior, Ecology, and Conservation*, 289–316.

19   successfully lobbied to have them hunted, since wolves occasionally kill livestock Idaho Legislative Wolf Oversight Committee, *Idaho Wolf Conservation and Management Plan* (2002).

20   Greenland, and began to build up on North America and northern Europe ed. Fagan, B. M., *The Complete Ice Age: How Climate Change Shaped the World* (London: Thames & Hudson, 2009).

20   fossil record on the other side of the world in North America Tedford, R. H., X. Wang, and B. E. Taylor, "Phylogenetic Systematics of the North American Fossil Caninae (*Carnivora: Canidae*)," *Bulletin of the American Museum of Natural History* 325, (2009): 1–218.

20   a small coyote, with a more robust build and a large head Miller, W. E., and O. Carranza-Castañeda, "Late Tertiary Canids from Central Mexico," *Journal of Paleontology* 72, no. 3 (1998): 546–56.

21   warm and temperate to freezing see Fagan, *The Complete Ice Age*.

21   Atlantic cool and isolated from warmer equatorial currents see *The IUCN Red List*.

21   Half the vegetation disappeared see Fagan, *The Complete Ice Age*.

21   interglacial periods, they returned to reoccupy their former habitats Wang, X., and R. H. Tedford, *Dogs: Their Fossil Relatives and Evolutionary History* (New York: Columbia University Press, 2008).

22   the Beringian land bridge connecting North America to Asia see Wang and Tedford, *Dogs*.

22   The Etruscan wolf was smaller than modern wolves Agustí, J., and M. Antón, *Mammoths, Sabertooths, and Hominids: 65 Million Years of Mammalian Evolution in Europe* (New York: Columbia University Press, 2002).

22   a skull similar to the American coyote see Wang and Tedford, *Dogs*.

22   conquest so complete, it became known as the Wolf Event Azzaroli, A., "Quaternary Mammals and the 'End-Villafranchian' Dispersal Event – A Turning Point in the History of Eurasia," *Palaeogeography, Palaeoclimatology, Palaeoecology* 44 (1983): 117–39.

22   its massive cheek teeth in a powerful skull see Agustí and Antón, *Mammoths, Sabertooths, and Hominids*.
        and, Turner, et al., "The Giant Hyaena, *Pachycrocuta brevirostris* (Mammalia, Carnivora, Hyaenidae)," *Geobios* 29, no. 4 (1996): 455–68

22 *Homo erectus* had large brains and fast-moving limbs Schrenk, F., and S. Müller, *The Neanderthals* (London, New York: Routledge, 2009).

22 palaeontologists found the remains of our ancestors *Homo erectus* Vekua, A., et al., "A New Skull of Early *Homo* from Dmanisi, Georgia," *Science* 297, no. 5578 (2002): 85–89.

23 Pacific Oceans, reaching all the way down to New York see Fagan, *The Complete Ice Age.*

23 twenty hours a day grazing up to 180 kilograms of grass Barton, M. et al., *Wild New World: Recreating Ice-Age North America* (London: BBC Books, 2002).

23–24 The hyena (*Crocuta crocuta*) was around 25 percent larger Churchill, S. *Thin on the Ground.* At press.

24 The leopard, *Panthera pardus,* was around the same size it is today in Africa Turner, A., *The Big Cats and Their Fossil Relatives* (New York: Columbia University Press, 1997).

 and, Leonard, J. A., et al., "Megafaunal Extinctions and the Disappearance of a Specialized Wolf Ecomorph," *Current Biology* 17, no. 13 (2007): 1146–50.

24 the saber-toothed cats were the top predators at the time Gonyea, W. J., "Behavioral Implications of Saber-toothed Felid Morphology," *Paleobiology* 2, no. 4 (1976): 332–42.

24 were capable of bringing down prey much larger than themselves Ibid.

24 the Neanderthals began 127,000 years ago see Schrenk and Müller, *The Neanderthals.*

25 the chilling air of the Ice Age before it reached their lungs see Ibid.

25 crack open bones for nutritious marrow see Churchill, *Thin on the Ground.*

25 Neanderthals had even bigger brains than modern humans Ponce de León, M. S., et al., "Neanderthal Brain Size at Birth Provides Insights into the Evolution of Human Life History," *Proceedings of the National Academy of Sciences* 105, no. 37 (2008), 13764–68.

25 larger Neanderthal population definitely went extinct Green, R. E., et al., "Analysis of One Million Base Pairs of Neanderthal DNA," *Nature* 444, no. 7117 (2006), 330–36.

 and, Yotova, V., et al., "An X-linked Haplotype of Neandertal Origin Is Present Among All Non-African Populations," *Molecular Biology and Evolution* 28, no. 7 (2011): 1957–62.

25 Some say it was climate change. Others say it was direct Gilligan, I., "Neanderthal Extinction and Modern Human Behaviour: The Role of Climate Change and Clothing," *World Archaeology* 39, no. 4 (2007): 499–514.

 and, Mellars, P., "Neanderthals and the Modern Human Colonization of Europe," *Nature* 432, no. 7016 (2004): 461–65.

 and, Horan, R. D., E. Bulte, and J. F. Shogren, "How trade saved humanity from Biological Exclusion: An economic theory of Neanderthal Extinction," *Journal of Economic Behavior & Organization* 58, no. 1 (2005): 1–29.

26 especially in the head and neck areas Berger, T. D., and E. Trinkaus. "Patterns of Trauma Among the Neandertals," *Journal of Archaeological Science* 22, no. 6 (1995): 841–52.

26 **a lot of unfriendly contact with large mammals** see Schrenk and Müller, *The Neanderthals.*

26 **overpower your competitors, and you have to be social** see Churchill, *Thin on the Ground.*

26 **Neanderthals lived in groups of only fifteen or so** see Schrenk and Müller, *The Neanderthals.*

27 **all the herbivores killed by predators** Karanth, K. U., et al., "Tigers and Their Prey: Predicting carnivore densities from Prey Abundance," *Proceedings of the National Academy of Sciences* 101, no. 14 (2004): 4854–58.

28 **the fossil record around a million years ago in Alaska** see Tedford, et al., "Phylogenetic Systematics."

28 **and they arrived in Europe approximately half a million years ago** Sotnikova, M., and L. Rook. "Dispersal of the Canini (Mammalia, Canidae: Caninae) Across Eurasia During the Late Miocene to Early Pleistocene," *Quaternary International* 212, no. 2 (2010): 86–97.

28 **humans intentionally adopted wolf puppies and tamed them on purpose** see Mech, "Canis Lupus."
    and, Clutton-Brock, "Man-made Dogs."
    and, Serpell and Paul, "Pets in the Family."

28 **Somewhere in early history a young wolf was brought into the family** see Mech, "Canis Lupus."

28 **our most successful and useful experiment in domestication** see Mech, *The Wolf.*

29 **And wolves eat a lot of meat** see Míklósi, *Dog Behaviour, Evolution, and Cognition.*

29 **as much as eleven pounds per wolf per day** Peterson, R. O., and P. Ciucci, "The Wolf as a Carnivore." In *Wolves: Behavior, ecology, and conservation,* 104–30.

29 **Starvation is a significant cause of mortality in many carnivores** Van Valkenburgh, B., "Iterative Evolution of Hypercarnivory in Canids (Mammalia: Carnivora): Evolutionary Interactions Among Sympatric Predators," *Paleobiology* 17, no. 4 (1991): 340–62.
    and, Palmqvist, P., A. Arribas, B. Martínez-Navarro, "Ecomorphological Study of Large Canids from the Lower Pleistocene of Southeastern Spain," *Lethaia* 32, no. 1 (1999): 75–88.

29 **As a general rule, it takes about twenty-two thousand pounds of prey** Carbone, C., and J. L. Gittleman, "A Common Rule for the Scaling of Carnivore Density," *Science* 295, no. 5563 (2002): 2273–76.

29 **Wolves are extremely possessive of their food** Koler-Matznick, J., "The Origin of the Dog Revisited," *Anthrozoös* 15, no. 2 (2002): 98–118.

29 **numerous bites until the prey falls** see Peterson and Ciucci, "The Wolf as a Carnivore."

29 **There are frequently squabbles over food** see Peterson and Ciucci, "The Wolf as a Carnivore."

30 **the hand gently lies on another skeleton – a puppy** see Morey, *Dogs.*
    and, Davis, S. J. M., and F. R. Valla, "Evidence for Domestication of the Dog 12,000 Years Ago in the Natufian of Israel," *Nature* 276, no. 608 (1978): 608–10.

30 **earliest records of humans being buried with another species – in this case a dog** see Davis and Valla, "Evidence for Domestication."

and, Bar-Yosef, O., "The Natufian Culture in the Levant, Threshold to the Origins of Agriculture," *Evolutionary Anthropology: Issues, News, and Reviews* 6, no. 5 (1998): 159–77.

30 **Similarly aged burials of dogs have been uncovered** see Morey, *Dogs*.

and, Davis and Valla, "Evidence for Domestication."

and, A number of research teams have used genetic techniques to establish where dogs were first domesticated (Larson, G., et al., "Rethinking Dog Domestication by Integrating Genetics, Archeology, and Biogeography," *Proceedings of the National Academy of Sciences* 109, no. 23 [2012]: 8878–83). Using mitochondrial DNA, it was first suggested that dogs were domesticated in East Asia (Savolainen, P., et al., "Genetic Evidence for an East Asian Origin of Domestic Dogs," *Science* 298, no. 5598 [2002]: 1610–13). Then, using nuclear DNA for analysis of single-nucleotide polymorphisms (SNPs), it was suggested that dogs originated in the Middle East (see vonHoldt, "Genome-wide SNP and Haplotype Analyses"). However, a recent review of the genetic, biogeographic, and archaeological data suggests that genetic techniques have failed to establish that dogs originate from a single region. Therefore it remains possible that dogs originated multiple times in several locations – likely across Eurasia. It is also possible that future genetic techniques that use full genome comparisons will successfully identify where dogs were first domesticated (see Larson, "Rethinking Dog Domestication").

31 **so strong that the two were often buried together** see Morey, *Dogs*.

Chapter 3: In My Parents' Garage (pages 33–61)

36 **our brain growth must occur after our birth** Rosenberg, K., and W. Trevathan, "Birth, Obstetrics and Human Evolution," *BJOG: An International Journal of Obstetrics and Gynaecology* 109, no. 11 (2002): 1199–206.

36 **Early skills become the foundation for more complex skills** Herrmann, E., et al., "Humans Have Evolved Specialized Skills of Social Cognition: The Cultural Intelligence Hypothesis," *Science* 317, no. 5843 (2007): 1360–66.

and, Herrmann, E., et al., "The Structure of Individual Differences in the Cognitive Abilities of Children and Chimpanzees," *Psychological Science* 21, no. 1 (2010): 102–10.

36 **develop powerful social skills as early as nine months old** Carpenter, M., et al., "Social Cognition, Joint Attention, and Communicative Competence from 9 to 15 Months of Age," *Monographs of the Society for Research in Child Development* 63, no. 4 (1998): 1–143.

and, Tomasello, M. "Joint Attention as Social Cognition." In *Joint Attention: Its Origins and Role in Development*, eds. C. Moore and P. J. Dunham (Psychology Press, 1995), 103–30.

37 **infants are beginning to read other people's intentions** Behne, T., M. Carpenter, and M. Tomasello, "One-Year-Olds Comprehend the Communicative

Intentions Behind Gestures in a Hiding Game," *Developmental Science* 8, no. 6 (2005): 492–99.

and, Tomasello, M., "Having Intentions, Understanding Intentions, and Understanding Communicative Intentions." In *Developing Theories of Intention: Social Understanding and Self-Control*, eds. P. D. Zelazo, J. W. Astington, and D. R. Olson (Psychology Press, 1999), 63–75.

and, Csibra, G., and G. Gergely, "Natural Pedagogy," *Trends in Cognitive Sciences* 13, no. 4 (2009): 148–53.

37 **all human forms of culture and communication** see Ibid.

and, Tomasello, M., et al., "Understanding and Sharing Intentions: The Origins of Cultural Cognition," *Behavioral and Brain Sciences* 28, no. 5 (2005): 675–90.

37 **problems learning language, imitating, and interacting with other people** Boria, S., et al., "Intention Understanding in Autism," *PLoS ONE* 4, no. 5 (2009), e5596.

and, Carpenter, M., M. Tomasello, and T. Striano, "Role Reversal Imitation and Language in Typically Developing Infants and Children with Autism" *Infancy* 8, no. 3 (2005): 253–78.

and, Pelphrey, K. A., J. P. Morris, and G. McCarthy, "Neural Basis of Eye Gaze Processing Deficits in Autism," *Brain* 128, no. 5 (2005): 1038–48.

and, Prothmann, A., C. Ettrich, and S. Prothmann, "Preference for, and Responsiveness to, People, Dogs and Objects in Children with Autism," *Anthrozoös* 22, no. 2 (2009): 161–71.

38 **unprecedented flexibility in solving problems** see Tomasello, "Having Intentions, Understanding Intentions."

and, Hare, B., and M. Tomasello, "Chimpanzees Are More Skilful in Competitive than in Cooperative Cognitive Tasks," *Animal Behaviour* 68, no. 3 (2004): 571–81.

38 **Mike compared humans with our closest living relatives, bonobos and chimpanzees** Prüfer, K., et al., "The Bonobo Genome Compared with the Chimpanzee and Human Genomes," *Nature* 486, no. 7407 (2012): 527–31.

38 **If we have a skill that bonobos and chimpanzees do not have** Wrangham, R., and D. Pilbeam, "African Apes as Time Machines." In *All Apes Great and Small*, Vol. 1: African Apes, eds. B. M. F., Galdikas, et al. (New York: Kluwer Academic, 2002): 5–17.

and Hare, B., "From Hominoid to Hominid Mind: What Changed and Why?" *Annual Review of Anthropology* 40, no. 1 (2011): 293–309.

38 **as skilled as young infants, Mike would know he was on the wrong track** see Tomasello, "Having Intentions, Understanding Intentions."

and, Tomasello, M., A. C. Kruger, and H. H. Ratner, "Cultural learning," *Behavioral and Brain Sciences* 16, no. 3 (1993): 495–511.

38 **bonobos and chimpanzees use visual gestures during social interactions** see Call, *The Gestural Communication*.

and, Liebal, K., and J. Call, "The Origins of Non Human Primates' Manual Gestures," *Philosophical Transactions of the Royal Society B: Biological Sciences* 367, no. 1585 (2012): 118–28.

and, Pollick, A. S., and F. B. M. de Waal, "Ape Gestures and Language Evolution," *Proceedings of the National Academy of Sciences* 104, no. 19 (2007): 8184–89.

and, Leavens, D. A., and W. D. Hopkins, "Intentional Communication by Chimpanzees: A Cross-Sectional Study of the Use of Referential Gestures," *Developmental Psychology* 34, no. 5 (1998), 813–22.

39   **a Scottish primatologist from the University of Stirling** Anderson, J. R., P. Sallaberry, and H. Barbier, "Use of Experimenter-Given Cues During Object-Choice Tasks by Capuchin Monkeys," *Animal Behaviour* 49, no. 1 (1995): 201–8.

39   **Mike thought they would do better than monkeys. But chimpanzees also failed** Call, J., B. A. Hare, and M. Tomasello, "Chimpanzee Gaze Following in an Object-Choice Task," *Animal Cognition* 1, no. 2 (1998): 89–99.

and, Agnetta, B., B. Hare, and M. Tomasello, "Cues to Food Location That Domestic Dogs (*Canis familiaris*) of Different Ages Do and Do Not Use," *Animal Cognition* 3, no. 2 (2000): 107–12.

39   **Even if they eventually learned they should choose the container you pointed at** Povinelli, D. J., et al., "Exploitation of Pointing as a Referential Gesture in Young Children, but not Adolescent Chimpanzees," *Cognitive Development* 12, no. 4 (1997): 423–61.

39   **The only exception was when the chimpanzees had been raised by humans** The only chimpanzees who have shown spontaneous skill at solving these same types of problems are individuals who were reared with intense exposure to humans (see Call, Hare, and Tomasello, "Chimpanzee Gaze Following"; and, Tomasello, M., and J. Call, "The Role of Humans in the Cognitive Development of Apes Revisited," *Animal Cognition* 7, no. 4 [2004]: 213–15; and, Call, J., B. Agnetta, and M. Tomasello, "Cues That Chimpanzees Do and Do Not Use to Find Hidden Objects," *Animal Cognition* 3, no. 1 [2000]: 23–34).

41   **see if they adapt their skills in flexible and intelligent ways** Tomasello, M., and J. Call, "Assessing the Validity of Ape-Human Comparisons: A Reply to Boesch (2007)," *Journal of Comparative Psychology* 122, no. 4 (2008): 449–52.

42   **except for the factors you wish to investigate** Martin, P. R., and P. Bateson, *Measuring Behaviour: An Introductory Guide* (Cambridge, UK: Cambridge University Press, 1993).

42   **he just got lucky and found a way to open it by accident** Morgan, C. L., *An Introduction to Comparative Psychology*, new edition, revised (London: Walter Scott Publishing, 1914).

42   **capuchin monkeys could make spontaneous inferences while using tools** Visalberghi, E., and L. Limongelli. "Lack of Comprehension of Cause, Effect Relations in Tool-Using Capuchin Monkeys (Cebus apella)," *Journal of Comparative Psychology* 108, no. 1 (1994): 15–22.

44   **I have no idea how it works, either** Ibid.

44   **these higher processes in the animal under observation** see Morgan, *An Introduction to Comparative Psychology*, 59.

44   **the latch problem immediately if they see someone else solve it first** Miller, H. C., R. Rayburn-Reeves, and T. R. Zentall, "Imitation and Emulation by

Dogs Using a Bidirectional Control Procedure," *Behavioural Processes* 80, no. 2 (2009): 109–14.

45 **Oreo as researchers used with infants, monkeys, and chimpanzees** see Hare, Call, and Tomasello, "Communication of Food Location."

45 **surreptitiously placed food under the other one** Because this was an experiment across trials within a session, we alternated in which cup the food was hidden, and we also alternated whether we really baited or fake baited each cup. We did not simply hide the food in the same cup repeatedly.

47 **thoroughly ruled out the possibility that dogs use their sense of smell to find the food in this context** This is always people's first question about our studies of dog cognition, and it was also ours. If dogs could simply find the food by smelling it from afar, we could not use games designed for infants to test dogs. So many scientists have examined the question and all found the same answer. On their first choice (which is the only one we consider in cognitive testing normally), dogs always perform at chance levels. More than a dozen studies from seven different research groups show that dogs, wolves, and foxes do not find food if hidden on their first choice in this context (see Hare, Call, and Tomasello, "Communication of Food Location"; and, Anderson, Sallaberry, and Barbier, "Use of Experimenter-Given Cues"; and, Agnetta, Hare, and Tomasello, "Cues to Food Location"; and, Hare, B., J. Call, and M. Tomasello, "Do Chimpanzees Know What Conspecifics Know?" *Animal Behaviour* 61, no. 1 [2001]: 139–51; and, Hare, B., et al., "Social Cognitive Evolution in Captive Foxes Is a Correlated By-product of Experimental Domestication," *Current Biology* 15, no. 3 [2005]: 226–30; and, Miklósi, A., et al., "Intentional Behaviour in Dog-Human Communication: An Experimental Analysis of 'Showing' Behaviour in the Dog," *Animal Cognition* 3, no. 3 [2000]: 159–66; and, McKinley, J., and T. D. Sambrook, "Use of Human-Given Cues by Domestic Dogs [*Canis familiaris*] and Horses [*Equus caballus*]," *Animal Cognition* 3, no. 1 [2000]: 13–22; and, Szetei, V., et al., "When Dogs Seem to Lose Their Nose: An Investigation on the Use of Visual and Olfactory Cues in Communicative Context Between Dog and Owner," *Applied Animal Behaviour Science* 83, no. 2 [2003]: 141–52; and, Bräuer, J., et al., "Making Inferences About the Location of Hidden Food: Social Dog, Causal Ape," *Journal of Comparative Psychology* 120, no. 1 [2006]: 38–47; and, Udell, M. A., R. F. Giglio, and C. D. L. Wynne, "Domestic Dogs [*Canis familiaris*] Use Human Gestures but Not Nonhuman Tokens to Find Hidden Food," *Journal of Comparative Psychology* 122, no. 1 [2008]: 84–93; and, Ittyerah, M., and F. Gaunet, "The Response of Guide Dogs and Pet Dogs [*Canis familiaris*] to Cues of Human Referential Communication [Pointing and Gaze]," *Animal Cognition* 12, no. 2 [2009]: 257–65; and, Hauser, M. D., et al., "What Experimental Experience Affects Dogs' Comprehension of Human Communicative Actions?" *Behavioural Processes* 86, no. 1 [2011]: 7–20; and, Riedel, J., et al., "Domestic Dogs [*Canis familiaris*] Use a Physical Marker to Locate Hidden Food," *Animal Cognition* 9, no. 1 [2006]: 27–35; and, Riedel, J., et al., "The Early Ontogeny of Human–Dog Communication," *Animal Behaviour* 75, no. 3 [2008]: 1003–14).

Even in tests where dogs are given dozens of trials and could potentially learn to smell the food, they do not (see control in Hare, B., et al., "The Domestication of Social Cognition in Dogs," *Science* 298, no. 5598 [2002]: 1634–36). Only a handful of individuals from the hundreds tested have performed above chance in this type of control. (This is again consistent with chance, since 5 percent should perform above chance based on probability; see Riedel, "Domestic Dogs [*Canis familiaris*] Use a Physical Marker"). Two additional controls have been run that further strengthen the argument against the use of smell in these types of tests. When dogs are shown food hidden in one cup but then, without them knowing, food is actually hidden in the opposite cup, dogs go to where they saw food hidden (88 percent) as opposed to where it is actually hidden (see Szetei, et al., "When Dogs Seem to Lose"). Finally, when someone points to a cup, and both cups have food (or smell like food because of steps taken to cover both locations with scent), dogs still use the pointing cue to find food (see Szetei, et al., "When Dogs Seem to Lose"; and, Hauser, "What Experimental Experience Affects"). The only context in which dogs are successful in using olfactory information is if they are allowed to closely inspect both hiding locations within a few centimetres and then brought back to start and allowed to choose (here they succeed). Alternatively, they are slightly above chance if an extremely smelly type of sausage is intentionally used to help them find the food using olfactory cues (60 percent success, see Szetei, et al., "When Dogs Seem to Lose").

47   **go *farther* away from my finger to get to it** see Riedel, et al., "The Early Ontogeny," demonstrating the same skill in puppies as young as six weeks of age.

and, Lakatos, G., et al., "Comprehension and Utilization of Pointing Gestures and Gazing in Dog–Human Communication in Relatively Complex Situations," *Animal Cognition* 15, no. 2 (2012): 201–13. Showing dogs using a pointing gesture to decide between four different hiding locations.

48   **he was nearly perfect from the start** Dogs use most human gestures on their first trial (for example, see Hauser, "What Experimental Experience Affects") and do not tend to improve in their performance during the experiment itself. Dogs make correct choices similar numbers of times in the first and second half of test sessions (see Hare, Call, and Tomasello, "Communication of Food Location"; and, Riedel, et al., "The Early Ontogeny"; and, Hare, et al., "The Domestication of Social Cognition"; and, Lakatos, G., et al., "A Comparative Approach to Dogs' [*Canis familiaris*] and Human Infants' Comprehension of Various Forms of Pointing Gestures," *Animal Cognition* 12, no. 4 [2009]: 621–31; and, Hare, B., and M. Tomasello, "Domestic Dogs [*Canis familiaris*] Use Human and Conspecific Social Cues to Locate Hidden Food," *Journal of Comparative Psychology* 113, no. 2 [1999]: 173–77).

48   **turned my head and looked at the cup** Subsequent studies have shown that dogs will respond to 'bowing' or bending at the waist thirty degrees towards the correct cup. They also use 'nodding' or a gaze towards the correct location followed by a nod as if indicating 'yes' (see Anderson, Sallaberry, and Barbier, "Use of Experimenter-Given Cues"; and, Udell, Giglio, and Wynne, "Domestic Dogs").

49   **stop pointing while the dog is choosing where to look** Miklósi, Á., and K. Soproni, "A Comparative Analysis of Animals' Understanding of the Human Pointing Gesture," *Animal Cognition* 9, no. 2 (2006): 81–93.

50   **The dogs followed my gaze and pointing gestures as skilfully as Oreo and Daisy** see Hare and Tomasello, "Domestic Dogs."

50   **Other research groups have since received similar results** see Anderson, Sallaberry, and Barbier, "Use of Experimenter-Given."

50   **Maggie's gaze and body orientation to find the food** see Hare and Tomasello, "Domestic Dogs."

51   **the motion created by placing the block** A piece of foam was used to occlude the dogs' view as the block was placed. All they saw was the experimenter holding the block before it went behind the occluder. Then the occluder was removed, and they could make a choice. Again, the dogs had a strong preference for choosing the cup with the block (see Agnetta, Hare, and Tomasello, "Cues to Food Location"). However, this particular motion control did not replicate in a recent study by another group (see Riedel, et al., "The Early Ontogeny").

52   **Dogs were better at understanding our gestures than chimpanzees** see Hare, et al., "The Domestication of Social Cognition in Dogs."

52   **the idea that dogs were doing something similar** see Call, Hare, and Tomasello, "Chimpanzee Gaze Following."

52   **dogs are selective and do not just use any kind of clue** Soproni, K., et al., "Comprehension of Human Communicative Signs in Pet Dogs (*Canis familiaris*)," *Journal of Comparative Psychology* 115, no. 2 (2001): 122–26.

53   **ignore other non-communicative clues** Soproni, K., et al., "Dogs' (*Canis familiaris*) Responsiveness to Human Pointing Gestures," *Journal of Comparative Psychology* 116, no. 1 (2002): 27–34.

      Dogs ignore cues given by inanimate objects. If someone extends a stick towards the cup, or a baby doll or plush toy animal is positioned with an arm extended towards a cup, dogs ignore this. Dogs also do not use human behaviours that are usually non-communicative. Dogs ignore shoulder wiggling, head tilting, or an elbow pointed towards the cup (see Udell, Giglio, and Wynne, "Domestic Dogs"; and, Hauser, et al., "What Experimental Experience Affects"). However, there are a couple of communicative cues that dogs typically fail to use. Very few dogs are spontaneously able to use a change in the direction of eye gaze if it is not paired with head movement in the same direction. They also fail to use a pointing cue if you place your hand on your belly button and only extend your index finger. Both of these communicative cues seem too subtle for most dogs to use reliably (see Hare, Call, and Tomasello, "Communication of Food Location"; and, Hare, Call, and Tomasello, "Do Chimpanzees Know"; and, Soproni, et al., "Comprehension of Human Communicative Signs").

53   **dogs do not use your gaze to find food** Trained chimpanzees can learn to follow human gaze to find hidden food, but they use a gaze cue regardless of whether the human looks at the cup or over the cup. In contrast, infants respond only to someone gazing at but not over a potential hiding place (see Soproni, et al., "Comprehension of Human Communicative Signs").

53 **at them before indicating where they should look** Téglás, E., et al., "Dogs' Gaze Following Is Tuned to Human Communicative Signals," *Current Biology* 22, no. 3 (2012): 209–12.

53 **alternate your gaze between them and the correct cup while pointing** see Bräuer, et al., "Making Inferences About."

and, Kupán, K., et al., "Why Do Dogs (*Canis familiaris*) Select the Empty Container in an Observational Learning Task?" *Animal Cognition* 14, no. 2 (2010), 259–68.

and, Hauser, et al., "What Experimental Experience."

but, see Kaminski, J., L. Schulz, and M. Tomasello, "How Dogs Know When Communication Is Intended for Them," *Developmental Science* 15, no. 2 (2011): 222–32.

53 **field site in Uganda** Wrangham, R., et al., "Chimpanzee Predation and the Ecology of Microbial Exchange," *Microbial Ecology in Health and Disease* 12, no. 3 (2000), 186–88.

54 **perhaps dogs raised by humans had learned the same skills** see Agnetta, Hare, and Tomasello, "Cues to Food Location That Domestic Dogs."

and, Hare, et al., "The Domestication of Social Cognition in Dogs."

and, Agnetta, Hare, and Tomasello, "Cues to Food Location That Domestic Dogs."

54 **exposure to humans, they still were performing nearly perfectly** see Hare, et al., "The Domestication of Social Cognition in Dogs."

54 **spontaneously use different types of human gestures** see Riedel, et al., "The Early Ontogeny."

and, Gácsi, M., et al., "Explaining Dog Wolf Differences in Utilizing Human Pointing Gestures: Selection for Synergistic Shifts in the Development of Some Social Skills," *PLoS ONE* 4, no. 8 (2009): e6584.

55 **more daily attention than other dogs** Gácsi, M., et al., "The Effect of Development and Individual Differences in Pointing Comprehension of Dogs," *Animal Cognition* 12, no. 3 (2009): 471–79.

55 **understanding a human pointing gesture** This study included dogs of a wide variety of breeds. However, the current study does not suggest that all dogs are the same – only that rearing history does not explain much, if any, of the observed variation in skill (but see debate about the impact of rearing in Udell, M. A. R., N. R. Dorey, and C. D. L. Wynne, "Wolves Outperform Dogs in Following Human Social Cues," *Animal Behaviour* 76, no. 6 [2008]: 1767–73; and, Wynne, C. D. L., M. A. R. Udell, and K. A. Lord, "Ontogeny's Impacts on Human-Dog Communication," *Animal Behaviour* 6, no. 8 [2008]: 1–6; and, Hare, B., et al., "The Domestication Hypothesis for Dogs' Skills with Human Communication: A Response to Udell et al. [2008] and Wynne et al. [2008]." *Animal Behaviour* 79, no. 2 [2010]: e1–e6).

55 **rescue dogs are just as skilled at using a variety of human social gestures as family-reared dogs** see Hare, et al., "The Domestication Hypothesis for Dogs' Skills."

but, see Barrera, G., A. Mustaca, and M. Bentosela, "Communication Between Domestic Dogs and Humans: Effects of Shelter Housing upon the Gaze to the Human," *Animal Cognition* 14, no. 5 (2011): 727–34.

and, Udell, M. A. R., N. R. Dorey, and C. D. L. Wynne, "The Performance of Stray Dogs (*Canis familiaris*) Living in a Shelter on Human-Guided Object-Choice Tasks," *Animal Behaviour* 79, no. 3 (2010): 717–25.

55 **human put the block on a cup, this was the cup they preferred** see Agnetta, Hare, and Tomasello, "Cues to Food Location That Domestic Dogs."
  and, Riedel, "Domestic Dogs (*Canis familiaris*) Use a Physical Marker."

55 **go away from the human to approach the correct cup** see Riedel, et al., "The Early Ontogeny."

55 **some breeds can be affected by training** see McKinley, "Use of Human-Given Cues by Domestic."

55 **and dogs can get better as they get older** see Udell, Giglio, and Wynne, "Domestic dogs (*Canis familiaris*) Use Human Gestures."
  and, Riedel, et al., "The Early Ontogeny."

56 **wolves should be as skilled at reading human communicative gestures as dogs** see Hare and Tomasello, "Domestic Dogs (*Canis familiaris*) Use Human."

56 **wolves reared by humans can outperform dogs on the exact same learning task** see Hare, et al., "The Domestication of Social Cognition."
  and, Frank, H., et al., "Motivation and Insight in Wolf (*Canis lupus*) and Alaskan Malamute (*Canis familiaris*): Visual Discrimination Learning." *Bulletin of the Psychonomic Society* 27, no. 5 (1989): 455–58.

56–57 **they had no interest in interacting with us** see Agnetta, Hare, and Tomasello, "Cues to Food Location That Domestic Dogs."

57 **attracted 30,000 visitors a year** For more on Wolf Hollow, please visit www.wolfhollowipswich.org.

57 **wolves had an unusual amount of exposure to humans** As puppies, all of the wolves were adopted into a human family at ten days old and only interacted with littermates and people until they were five weeks old. At five weeks, they were placed in a holding pen with their mother so that they could continue to interact with people daily. At twelve weeks of age, the wolves were reintroduced to the entire pack; human caretakers interacted with them daily and could even enter their pen for management purposes (see Hare, et al., "The Domestication of Social Cognition").

58 **gestures to show them which cup hid the food, the wolves chose randomly** Ibid.

58 **puppies I had tested showed more skill at reading human gestures than wolves** Now we know that even six-week-old puppies perform better than these wolves with human gestures (Gácsi, M., et al., "Species Specific Differences and Similarities in the Behavior of Hand-Raised Dog and Wolf Pups in Social Situations with Humans," *Developmental Psychobiology* 47, no. 2 (2005): 111–22; and, Riedel, et al., "The Early Ontogeny").

58 **disinterested in food games or were anxious around humans** see Hare, et al., "The Domestication of Social Cognition."

59 **They needed explicit training to match the spontaneous performance of dog puppies** Gácsi, M. et al., "Explaining Dog Wolf Differences."
  and, Virányi, Z., et al., "Comprehension of Human Pointing Gestures in Young Human-Reared Wolves (*Canis lupus*) and Dogs (*Canis familiaris*)," *Animal Cognition* 11, no. 3 (2008), 373–87.

59　**demonstrate this skill without any training** One study has suggested a group of wolves outperformed a group of rescue dogs with a pointing cue (see Udell, Dorey, and Wynne, "Wolves Outperform Dogs"), but reanalysis of the data shows that dogs and wolves perform the same. This replicates previous findings with trained, heavily socialized wolves (see Gácsi, et al., "Explaining Dog Wolf Differences"; and, Hare, et al., "The Domestication Hypothesis"). The most interesting wolf data comes from a comparison of a group of heavily socialized eight-week-old wolf pups compared with same-aged dog pups (see Gácsi, et al., "Explaining Dog Wolf Differences"). While several of the wolf pups could not even be tested because the experimenters could not hold them, as a group they did perform above chance with a human pointing cue in the food-finding game. Since these socialized wolf pups show skill at the level of some dog puppies and the control foxes I tested, it would be fascinating to follow these individuals as they develop. The domestication hypothesis would predict that the wolves' skills at using human gestures would decrease as their adult fear system came online. These wolves should perform as poorly as previously tested socialized wolves of at least four months of age (see Ibid.). It would also be important to compare these wolves' use of a variety of human gestures, not just the pointing gesture, since it is the flexibility of our dogs' use of gestures that is remarkable.

59　**dogs' dependence on our social information** Topál, J., et al., "Differential Sensitivity to Human Communication in Dogs, Wolves, and Human Infants," *Science* 325, no. 5945 (2009): 1269–72.

59　**relying too heavily on humans can get dogs confused in some situations** see Ibid.

61　**bonobos and how difficult it was to explain their evolution** Wrangham, R. W., and D. Peterson, *Demonic Males: Apes and the Evolution of Human Aggression* (New York: Houghton Mifflin, 1996).

61　**It's in a chapter by Ray Coppinger** Trut, L. N., "Early Canid Domestication: The Farm-Fox Experiment," *American Scientist* 87, no. 2 (1999): 160–69.
　　　and, Coppinger, R., and R. Schneider, "Evolution of Working Dogs." In *The Domestic Dog: Its Evolution, Behaviour, and Interactions with People*, ed. J. Serpell (Cambridge, UK: Cambridge University Press, 1995), 21–47.

Chapter 4: Clever as a Fox (pages 63–94)

63　**the worst man-made famine in history** Kuromiya, H., *Stalin: Profiles in Power* (Harlow, UK, and New York: Pearson/Longman, 2005).

64　**from one generation to the next, he was not sure how** Zimmer, C., *Evolution: The Triumph of an Idea* (New York: Harper, 2001).

64　**found in Darwin's library after he died that could have solved everything** Henig, R. M. *The Monk in the Garden: The Lost and Found Genius of Gregor Mendel, the Father of Genetics* (New York: Houghton Mifflin, 2000).

64　**Darwin or the theory of natural selection** de Beer, G., "Mendel, Darwin, and Fisher (1865–1965)," *Notes and Records of the Royal Society of London* 19, no. 2 (1964), 192–226.

67   Darwin's copy was still uncut, which means he never read it see de Beer, "Mendel, Darwin, and Fisher."

67   Darwin together in what became a grand synthesis see Zimmer, *Evolution.*

67   then got a whole lot of other traits by accident Bidau, C., "Domestication through the centuries: Darwin's ideas and Dmitry Belyaev's long-term experiment in silver foxes," *Gayana* 73 (2009): 55–72.

67   winter has passed and it is safe to flower in the spring Medvedev, Z. A., *The Rise and Fall of T. D. Lysenko*, trans. I. M. Lerner (New York: Columbia University Press, 1969).

68   more lush and fertile and would save millions from starvation see Ibid.

68   country's wheat production up to tenfold Soyfer, V. N., "The Consequences of Political Dictatorship for Russian Science," *Nature Reviews Genetics* 2, no. 9 (2001): 723–29.

68   Lysenko began to falsify his results see Ibid.
       and, Grant, J., *Corrupted Science: Fraud, Ideology and Politics in Science* (Facts, Figures & Fun, 2007).

68   and eventually rejected the concept of the gene altogether see Soyfer, "The Consequences of Political."

68   American geneticist and Nobel laureate Hermann Muller see Ibid.

69   people arrested and half of those sentenced to death see Kuromiya, *Stalin.*

69   The Nazis thought of Russians as subhuman see Ibid.

69   justify the racism that existed in American society see Medvedev, *The Rise and Fall of T. D. Lysenko.*

69   Monuments were erected in his honour see Ibid.

69–70 Lysenkoites were almost completely uneducated Gershenson, S., "Difficult Years in Soviet Genetics," *Quarterly Review of Biology* 65, no. 4 (1990): 447–56.

70   genetics to be completely prohibited in the USSR see Soyfer, "The Consequences of Political."
       and, Gershenson, "Difficult Years in Soviet."
       and, Medvedev, *The Rise and Fall of T. D. Lysenko.*

70   colleagues, and it is impossible to obtain a copy Argutinskaya, S., "In Memory of D. K. Belyaev. Dmitrii Konstantinovich Belyaev: A Book of Reminescences (V. K. Shumnyi, P. M. Borodin, A. L. Markel', and S. V. Argutinskaya, eds., Novosibirsk: Sib. Otd. Ros. Akad. Nauk, 2002)," *Russian Journal of Genetics* 39, no. 7 (2003): 842–43.

70   Nikolai went on to become a geneticist Trut, L. N., et al., "To the 90th Anniversary of Academician Dmitry Konstantinovich Belyaev (1917– 1985)," *Russian Journal of Genetics* 43, no. 7 (2007): 717–20.

71   Nikolai was arrested by the secret police see Argutinskaya, "In Memory of D. K. Belyaev."

71   essentially beginning his career in genetics see Kuromiya, *Stalin.*

71   how incredibly courageous Belyaev was to continue his work after the war see Ibid.

71   Stalin had executed every army commander of the first and second rank see Ibid.

71 **at its highest – more than 2.5 million** Applebaum, A., *Gulag: A History* (New York: Doubleday, 2003).

71 **'Silver-Coloured Fur in Silver-Black Foxes'** see Bidau, "Domestication Through the Centuries."

71 **In 1948, when genetics was banned altogether, Belyaev was fired** see Trut, "Early Canid Domestication."

72 **Stalin was openly supportive of Lysenko** see Medvedev, *The Rise and Fall of T. D. Lysenko.*

72 **you had to do it in a chemistry, mathematics, or physics journal** see Ibid.

72 **'an experiment on a gigantic scale'** see Darwin, *The Variation of Animals and Plants under Domestication.*

72 **the struggle for survival drove evolution** see Bidau, "Domestication Through the Centuries."

73 **unmolested by the authorities, until his death in 1985** see Trut, "Early Canid Domestication."

73 **snowy shades that were the ultimate luxury item** Martin, J., *Treasure of the Land of Darkness: The Fur Trade and Its Significance for Medieval Russia* (Cambridge, UK: Cambridge University Press, 2004).

73 **Silver foxes are a variety of the red fox (*Vulpes vulpes*)** Baker, P. J., and Harris, S. "Red foxes: The Behavioural Ecology of Red Foxes in Urban Bristol." In *The Biology and Conservation of Wild Canids*, eds. D. W. Macdonald and C. Sillero-Zubiri (Oxford, UK: Oxford University Press, 2004), 207–16.

73 **responsible for the silver appearance** Cross, E. C., "Colour Phases of the Red Fox (*Vulpes fulva*) in Ontario," *Journal of Mammalogy* 22, no. 1 (1941): 25–39.

73 **bred on fur farms on Russia since the late nineteenth century** Belyaev, D., "Domestication of Animals," *Science Journal* 5, no. 1 (1969): 47–52.

73 **the silver quality of the coat** see Ibid.

73 **bite if you attempted to handle them** This same phenomenon has been observed in captive cavies, which are the species from which guinea pigs were domesticated. Although kept in captivity for thirty generations, captive cavies remain aggressive with one another and shy of humans. Life in captivity does not create a population of fearless and friendly cavies; only breeding individuals with a genetic predisposition for tameness will do so (Künzl, C., et al., "Is a Wild Mammal Kept and Reared in Captivity Still a Wild Animal?" *Hormones and Behavior* 43, no. 1 [2003]: 187–96).

74 **animals could reproduce at various times of the year** see Trut, "Early Canid Domestication."

74 **vixens from a fur farm in Estonia** see Ibid.

74 **and bite precautions had to be taken** see Belyaev, "Domestication of Animals."

74 **Belyaev called his initial fox population 'virtually wild animals'** Belyaev, D., "Destabilizing Selection as a Factor in Domestication," *Journal of Heredity* 70, no. 5 (1979): 301–8.

75 **no foreigner had ever collected and published data on the foxes** see Applebaum, *Gulag*.

76 **academics who shared scientific information could be criminally prosecuted** see Ibid.

77 **the foxes uncontrollably wagged their tails** Trut, L. N., et al., "Morphology and Behavior: Are They Coupled at the Genome Level?" In *The Dog and Its Genome*, eds. E. A. Ostrander, U. Giger, and K. Lindblad-Toh (Woodbury, NY: Cold Spring Harbor Laboratory Press, 2006): 81–93.

　　and, Kukekova, A. V., et al., "Mapping Loci for Fox Domestication: Deconstruction/Reconstruction of a Behavioral Phenotype," *Behavior Genetics* 41, no. 4 (2011): 593–606.

　　and, Gogoleva S. S., et al., "The Sustainable Effect of Selection for Behaviour on Vocalization in the Silver Fox," *VOGiS Herald* 12 (2008): 24–31.

　　and, Gogoleva, S. S., et al., "Vocalization toward Conspecifics in Silver Foxes (*Vulpes vulpes*) Selected for Tame or Aggressive Behavior toward Humans," *Behavioural Processes* 84, no. 2 (2010): 547–54.

77 **since they still have their parents' genes** Daniels, T. J., and M. Bekoff, "Feralization: The Making of Wild Domestic Animals," *Behavioural Processes* 19, nos. 1–3 (1989): 79–94.

　　and, Fox, M. W., *The Dog: Its Domestication and Behavior* (New York: Garland Publishing, 1978).

77 **changes that can be passed to the next generation** see Trut, "Early Canid Domestication"

77 **both populations had very little contact with humans** see Ibid.

77 **either group saw a human was when they were fed** see Hare, et al., "Social Cognitive Evolution."

77 **Yet just like puppies, these foxes showed affection towards humans** see Ibid.

78 **change caused by Belyaev's selection regime** see Trut, Oskina, and Kharlamova, "Animal Evolution during Domestication."

78 **a quarter of those seen among the control foxes** see Trut, "Early Canid Domestication."

　　and, Gulevich, R. G., et al., "Effect of Selection for Behavior on Pituitary-Adrenal Axis and Proopiomelanocortin Gene Expression in Silver Foxes (*Vulpes vulpes*)," *Physiology & Behavior* 82, no. 2 (2004): 513–18.

78 **neurotransmitter that makes humans feel happy and relaxed** see Trut, "Early Canid Domestication."

　　and, Popova, N., et al., "Effect of Domestication of the Silver Fox on the Main Enzymes of Serotonin Metabolism and Serotonin Receptors," *Genetika* 33, no. 3 (1997): 370–4.

78 **longer breeding season than the control foxes** see Trut, Oskina, and Kharlamova, "Animal Evolution During Domestication."

78 **differences observed between dogs and wolves** see Trut, "Early Canid Domestication."

　　and, Trut, et al., "Morphology and Behavior."

78 **who were less aggressive and more social towards people** see Trut, Oskina, and Kharlamova, "Animal Evolution During Domestication."

78 **domesticated animals and their wild ancestors** see Trut, "Early Canid Domestication."

79 **human-gesture-reading games than the control foxes** see Hare, et al., "Social Cognitive Evolution."

79 **foxes would fail to understand human gestures** see Hare, et al., "The Domestication of Social Cognition."
   and, Hare, et al., "Social Cognitive Evolution."

85 **touching one of the tape measure strips** When we compared the number of test trials in which the foxes touched one of the toys, there was no difference between the experimental and control populations. The control foxes made as many choices as the experimental foxes in both the gesture and feather test. Any difference we saw in their preference of which toy to touch could not be a result of the control foxes' being too afraid to make choices (see Ibid.).

86 **toy that the feather on a stick touched** When directly compared, the experimental foxes preferred the toy I had gestured towards more than the control foxes. However, it is not that the control foxes avoided the toy I touched. Instead, they just chose randomly. When we directly compared the two populations' choices in the feather test, the control foxes chose the toy touched by the feather more often than the experimental foxes. The experimental foxes did not avoid the toy touched by the feather. They just chose randomly, as did the control foxes in the gesture test (see Ibid.).

87 **When fear was replaced by curiosity about humans and toys** Although our fox kits were not afraid of humans, due to their socialization with humans, as they grew older they would have quickly had difficulty interacting with humans as their adult fear responses developed. This same type of fear response does not develop in the experimental foxes.

88 **intense socialization with humans** see Hare, et al., "Social Cognitive Evolution."

88 **"control its reproduction and (in the case of animals) its food supply"** Diamond, J., "Evolution, Consequences and Future of Plant and Animal Domestication," *Nature* 418, no. 6898 (2002): 700–707.

89 **Ray Coppinger and others** Darcy Morey also articulated this idea in a 1994 paper in *American Scientist*. See Morey, D. F., "The Early Evolution of the Domestic Dog: Animal Domestication, Commonly Considered a Human Innovation Can Also Be Described as an Evolutionary Process," *American Scientist* 82, no. 4 (1994): 140–51.
   and, Zeuner, F. E., *A History of Domesticated Animals* (London: Hutchinson and Co., 1963).

89 **rotting meat, and starchy vegetable scraps** Salvador, A., and P. Abad, "Food Habits of a Wolf Population (*Canis lupus*) in León Province, Spain," *Mammalia* 51, no. 1 (1987): 45–52.
   and, Boitani, L., "Wolf Research and Conservation in Italy," *Biological Conservation* 61, no. 2 (1992): 125–32.

and, Voigt, D. R., G. B. Kolenosky, and D. H. Pimlott, "Changes in Summer Foods of Wolves in Central Ontario," *Journal of Wildlife Management* 40, no. 4 (1976): 663–68.

and, Crete, M., R. Taylor, and P. Jordan, "Simulating Conditions for the Regulation of a Moose Population by Wolves," *Ecological Modelling* 12, no. 4 (1981): 245–52.

and, Mech, L. D., "The Challenge and Opportunity of Recovering Wolf Populations," *Conservation Biology* 9, no. 2 (1995): 270–78.

and, Beaver, B. V., M. Fischer, and C. E. Atkinson. "Determination of Favorite Components of Garbage by Dogs," *Applied Animal Behaviour Science* 34, nos. 1–2 (1992): 129–36.

89 **wolves to undergo physical changes** Species can evolve very rapidly under the right conditions. For example, one or two species of cichlid fish invaded Lake Victoria in East Africa one million to two million years ago; they have now exploded into five hundred separate species (see Kocher, T. D., "Adaptive Evolution and Explosive Speciation: The Cichlid Fish Model," *Nature Reviews Genetics* 5, no. 4 [2004]: 288–98).

89 **changes in coat color were already occurring in the eighth generation of foxes** see Trut, Oskina, and Kharlamova, "Animal Evolution during Domestication."

90 **taken as pets when young** This type of relationship still exists between people and 'village dogs' in many developing countries, such as Ethiopia (see Ortolani, A., H. Vernooij, and R. Coppinger, "Ethiopian Village Dogs: Behavioural Responses to a Stranger's Approach," *Applied Animal Behaviour Science* 119, nos. 3–4 [2009]: 210–18; and, Pal, S. K., "Maturation and Development of social Behaviour during Early Ontogeny in Free-Ranging Dog Puppies in West Bengal, India," *Applied Animal Behaviour Science* 111, nos. 1–2 [2008]: 95–107).

90 **What we needed was a modern version of these early dogs** Alternatively, it might be that only domestic dogs intentionally bred to be the pets of humans within the last two hundred years will show these unusual skills (see Parker, H. G., et al., "Genetic Structure of the Purebred Domestic Dog," *Science* 304, no. 5674 (2004): 1160–64). This would not support the idea that dogs domesticated themselves, since these traits should be linked as observed in the experimental foxes.

90 **domesticated dogs who have gone wild** Spotte, S., *Societies of Wolves and Free-Ranging Dogs* (Cambridge, UK: Cambridge University Press, 2012).

90 **that they do not have a human family** see Daniels and Bekoff, "Feralization."

90 **scavenging near human settlements** Boitani, L., and P. Ciucci, "Comparative Social Ecology of Feral Dogs and Wolves," *Ethology Ecology & Evolution* 7 (1995): 49–72.

91 **closely related to Asian breeds** Savolainen, P., et al., "A Detailed Picture of the Origin of the Australian Dingo, Obtained from the Study of Mitochondrial DNA," *Proceedings of the National Academy of Sciences* 101, no. 33 (2004), 12387–90.

91 **these dogs probably went feral as far back as five thousand years ago** see Koler-Matznick, "The Origin of the Dog Revisited."

and, Koler-Matznick, J., et al., "An Updated Description of the New Guinea Singing Dog (*Canis hallstromi*, Troughton 1957)," *Journal of Zoology* 261 (2003): 109–18.

and, Vilà, C., et al., "Multiple and Ancient Origins of the Domestic Dog," *Science* 276, no. 5319 (1997): 1687–89.

and, Purcell, B., *Dingo* (Collingwood: Csiro Publishing, 2010).

91 **researchers suspect that neither was intentionally bred by humans** see Koler-Matznick, et al., "An Updated Description."

and, Savolainen, et al., "A Detailed Picture of the Origin."

and, Smith, B. P., and C. A. Litchfield, "A Review of the Relationship Between Indigenous Australians, Dingoes (*Canis dingo*) and Domestic Dogs (*Canis familiaris*)," *Anthrozoös* 22, no. 2 (2009): 111–28.

91 **closest modern representative of proto-dogs** see Spotte, *Societies of Wolves.*

91 **prey they occasionally steal from harpy eagles** see Koler-Matznick, et al., "An Updated Description of the New Guinea Singing Dog."

91 **They are also one of the rarest dogs in the world** see Ibid.

92 **reading our gestures, even though no humans had bred them for this purpose** see Wobber, V., et al., "Breed Differences in Domestic Dogs' (*Canis familiaris*) Comprehension of Human Communicative Signals." *Interaction Studies* 10, no. 2 (2009): 206–24.

92 **dingoes have shown similar skills** Smith, B. P., and C. A. Litchfield, "Dingoes (*Canis dingo*) Can Use Human Social Cues to Locate Hidden Food," *Animal Cognition* 13, no. 2 (2010): 367–76.

93 **may also have been a crucial source of food** Clutton-Brock, J., and N. Hammond, "Hot dogs: Comestible Canids in Preclassic Maya Culture at Cuello, Belize." *Journal of Archaeological Science* 21, no. 6 (1994): 819–26.

93 **Like ravens following wolves to a kill** Heinrich, B., *Mind of the Raven: Investigations and Adventures with Wolf-Birds* (New York: Perennial, 2007).

93 **honeyguide birds to beehives full of honey** Marlowe, F., *The Hadza: Hunter-Gatherers of Tanzania*, vol. 3 (Berkeley: University of California Press, 2010).

and, Wrangham, R. *Honey and Fire in Human Evolution.* At press.

93 **when the dogs cornered their prey** Koster, J., "The Impact of Hunting with Dogs on Wildlife Harvests in the Bosawas Reserve, Nicaragua," *Environmental Conservation* 35, no. 3 (2008): 211–20.

and, Koster, J. M., "Hunting with Dogs in Nicaragua: An Optimal Foraging Approach," *Current Anthropology* 49, no. 5 (2008): 935–44.

93 **humans prefer other wolves over humans** see Gácsi, et al., "Species-Specific Differences."

and, Topál, J., et al., "Attachment Behavior in Dogs (*Canis familiaris*): A New Application of Ainsworth's (1969) Strange Situation Test," *Journal of Comparative Psychology* 112, no. 3 (1998): 219–29.

and, Topál, J., et al., "Attachment to Humans: A Comparative Study on Hand-Reared Wolves and Differently Socialized Dog Puppies," *Animal Behaviour* 70 (2005): 1367–75.

93 **people survived to produce the next generation** Corballis, M. C., and S. E. G. Lea, *The Descent of Mind*: Psychological Perspectives on Hominid Evolution (Oxford, UK: Oxford University Press, 1999).

and, Byrne, R., and A. Whiten, *Machiavellian Intelligence: Social Expertise and the Evolution of Intellect in Monkeys, Apes, and Humans* (New York: Oxford University Press, 1989).

and, Barkow, J. H., L. Cosmides, and J. Tooby, *The Adapted Mind: Evolutionary Psychology and the Generation of Culture* (New York: Oxford University Press, 1995).

Chapter 5: Survival of the Friendliest (pages 95–121)

96 **originated somewhere in the Niger-Congo region** Johannes, J., "Basenji Origin and Migration: Into the Heart of Africa," *Official Bulletin of the Basenji Club of America* 39, no. 4 (2005): 60–62.

96 **Genetically, basenjis are among nine living breeds that are more wolf-like** see Parker, "Genetic Structure."

96 **most beautiful sight I had ever seen** The Congo Basin in the Democratic Republic of the Congo (DRC) remains almost completely untouched in comparison with other neighbouring countries. Satellite imagery suggests as much as 98 percent of the forest in DRC is still virgin (see Laporte, N. T., et al., "Expansion of Industrial Logging in Central Africa," *Science* 316, no. 5830 [2007]: 1451). This makes these forests a conservation high priority since they act as a carbon sink crucial to moderating global warming (see Gibbs, H. K., et al., "Monitoring and Estimating Tropical Forest Carbon Stocks: Making REDD a Reality," *Environmental Research Letters* 2, 045023 (2007).

97 **African continent that became the Congo Basin** Myers Thompson, J., "A Model of the Biogeographical Journey from Proto pan to Pan Paniscus," *Primates* 44, no. 2 (2003): 191–97.

98 **some would exist nowhere else in the world** see Ibid.

98 **conservationist Claudine André** Claudine has won several prestigious awards for her efforts, including the highest civilian awards from both Belgium and France. A full-length feature movie was released in France in 2011 about her efforts to save orphan bonobos and release them back into the wild.

98 **adept at surviving all kinds of turbulence** André, C., *Une Tendresse Sauvage* (Paris: Calmann-Lévy, 2006).

99 **scientists who had advised on the process** André, C., et al., "The Conservation Value of Lola ya Bonobo Sanctuary." In *The Bonobos: Behavior, Ecology, and Conservation*, eds. T. Furuishi and J. Thompson (New York: Springer, 2008), 303–22.

100 **chimpanzees slowly took over parts of Kahama territory** Part of the horror stemmed from the fact that many of the individuals hunted down and killed by the Kasakela community had previously belonged to this community. Kasakela males were motivated to kill their former group mates after they became part of a new group (see Goodall, *The Chimpanzees of Gombe*).

100 **chimpanzees have a dark side** see Ibid.

101 **subsequently take over their territories** Wrangham, R. W., M. L. Wilson, and M. N. Muller, "Comparative Rates of Violence in Chimpanzees and Humans," *Primates* 47, no. 1 (2006): 14–26.

and, Mitani, J. C., and D. P. Watts, "Correlates of Territorial Boundary Patrol Behaviour in Wild Chimpanzees," *Animal Behaviour* 70, no. 5 (2005): 1079–86.

and, Wilson, M. L., and R. W. Wrangham, "Intergroup Relations in Chimpanzees," *Annual Review of Anthropology* 32 (2003): 363–92.

and, Mitani, J. C., D. P. Watts, and S. J. Amsler, "Lethal Intergroup Aggression Leads to Territorial Expansion in Wild Chimpanzees," *Current Biology* 20, no. 12 (2010): R507–R508.

and, The most carefully described case occurred when the Ngogo chimpanzee community in Kibale National Forest, Uganda, systematically killed twenty-eight members in a neighbouring group over a ten-year period. At the end of this period, they had absorbed into their own territory a large portion of land that had previously been in possession of the neighbours they had killed.

101 **These gangs typically target males and infants** Muller, M. N., "Chimpanzee Violence: Femmes Fatales," *Current Biology* 17, no. 10 (2007): R365–R366.

101 **levels in some pre-agricultural human societies** see Wrangham, Wilson, and Muller, "Comparative Rates of Violence."

101 **through the male hierarchy** see Goodall, *The Chimpanzees of Gombe.*

101 **alpha males is sexual control over the females** Boesch, C., and H. Boesch-Achermann, *The Chimpanzees of the Taï Forest: Behavioural Ecology and Evolution* (New York: Oxford University Press, 2000).

101 **Unfortunately, males often force females to mate with them** Muller, M. N., et al., "Male Coercion and the Costs of Promiscuous Mating for Female Chimpanzees," *Proceedings of the Royal Society B: Biological Sciences* 274, no. 1612 (2007): 1009–14.

101 **Genetically, bonobos are almost identical to chimpanzees** De Waal, F. B. M., and F. Lanting, *Bonobo: The Forgotten Ape* (Berkeley: University of California Press, 1997).

and, Prüfer, K., et al., "The Bonobo Genome Compared."

101 **another group's territory, or killing other bonobos** Kano, T., *The Last Ape: Pygmy Chimpanzee Behavior and Ecology* (Ann Arbor: University Microfilms, 1992).

and, Furuichi, T., "Female Contributions to the Peaceful Nature of Bonobo Society," *Evolutionary Anthropology: Issues, News, and Reviews* 20, no. 4 (2011): 131–42.

and, Gerloff, U., et al., "Intracommunity Relationships, Dispersal Pattern and Paternity Success in a Wild Living Community of Bonobos (*Pan paniscus*) Determined from DNA Analysis of Faecal Samples," *Proceedings of the Royal Society of London. Series B: Biological Sciences* 266, no. 1424 (1999): 1189–95.

and, Hohmann, G., and B. Fruth, "Intra- and Inter-Sexual Aggression by Bonobos in the Context of Mating," *Behaviour* 140, no. 11/12 (2003): 1389–413.

101 **even travel together for days as one giant bonobo group** see Kano, *The Last Ape.*

and, Furuichi, "Female Contributions to the Peaceful Nature of Bonobo Society."

and Surbeck, M., R. Mundry, and G. Hohmann, "Mothers Matter!: Maternal Support, Dominance Status and Mating Success in Male Bonobos (*Pan paniscus*)," *Proceedings of the Royal Society B: Biological Sciences* 278, no. 1705 (2011): 590–98.

102 **Hal Coolidge was riffling through** Hal Coolidge was founding director of both the World Wildlife Fund (WWF) and the International Union for the Conservation of Nature and Natural Resources (IUCN). Today, these are two of the main international organizations that protect endangered species and natural places all over the world.

102 **last of the great apes to be discovered** Coolidge, H., "*Pan paniscus*: Pygmy Chimpanzee from South of the Congo River," *American Journal of Physical Anthropology* 18, no. 1 (1933): 1–57.

102 **smaller than the cranium of a male chimpanzee** Cramer, D. L., "Craniofacial Morphology of *Pan paniscus*: A Morphometric and Evolutionary Appraisal," *Contributions to Primatology* 10 (1977): 1.

102–3 **Bonobo skulls are 'frozen' in a juvenile state** Shea, B. T., "Paedomorphosis and Neoteny in the Pygmy Chimpanzee," *Science* 222, no. 4623 (1983): 521–22.

and, Lieberman, D. E., et al., "A Geometric Morphometric Analysis of Heterochrony in the Cranium of Chimpanzees and Bonobos," *Journal of Human Evolution* 52, no. 6 (2007): 647–62.

and, Durrleman, S., et al., "Comparison of the Endocranial Ontogenies Between Chimpanzees and Bonobos Via Temporal Regression and Spatiotemporal Registration," *Journal of Human Evolution* 62, no. 1 (2011): 74–88

103 **domesticated fowl show a similar pattern** Kruska, D. C. T., "On the Evolutionary Significance of Encephalization in some Eutherian Mammals: Effects of Adaptive Radiation, Domestication, and Feralization," *Brain, Behavior and Evolution* 65, no. 2 (2005): 73–108.

and, Wayne, R. K., "Consequences of Domestication: Morphological Diversity of the Dog." In E. Ostrander and A. Ruvinsky, eds., *The Genetics of the Dog* (Oxfordshire: CABI, 2002), 43–60.

103 **usually along territorial boundaries of two different packs** Mech, L., et al., *The Wolves of Denali* (Minneapolis: University of Minnesota Press, 1998).

103 **competing for sexually receptive females** Derix, R., et al., "Male and Female Mating Competition in Wolves: Female Suppression vs. Male Intervention," *Behaviour* 127, nos. 1/2 (1993): 141–74.

103 **occasionally kill their puppies** McLeod, P., "Infanticide by Female Wolves," *Canadian Journal of Zoology* 68, no. 2 (1990): 402–4.

103 **nearly extinguished within a group's territory** Lwanga, J., et al., "Primate Population Dynamics over 32.9 Years at Ngogo, Kibale National Park, Uganda," *American Journal of Primatology* 73, no. 10 (2011): 997-1011.

103   **the puppies of other pack members** see Boitani and Ciucci, "Comparative Social Ecology."

and, Pal, S. K., B. Ghosh, and S. Roy, "Agonistic Behaviour of Free-Ranging Dogs (*Canis familiaris*) in Relation to Season, Sex and Age," *Applied Animal Behaviour Science* 59, no. 4 (1998): 331–48.

and, Macdonald, D. W., and G. M. Carr, "Variation in Dog Society: Between Resource Dispersion and Social Flux." In *The Domestic Dog*, 199–216.

and, Bonanni, R., et al., "Free-Ranging Dogs Assess the Quantity of Opponents in Intergroup Conflicts," *Animal Cognition* 14, no. 1 (2011): 103–15.

and, Pal, S. K., "Parental Care in Free-Ranging Dogs, *Canis familiaris*," *Applied Animal Behaviour Science* 90, no. 1 (2005): 31–47.

104   **even from their own fellow pack members** Bradshaw, J. W. S., and H. M. R. Nott, "Social and Communication Behavior of Companion Dogs." In *The Domestic Dog*, 115–30.

104   **juvenile wolves throughout their lives** see Míklósi, *Dog Behaviour, Evolution, and Cognition*.

and, Koler-Matznick, et al., "An Updated Description of the New Guinea Singing Dog."

and, Pal, S. K., "Play behaviour During Early Ontogeny in Free-Ranging Dogs (*Canis familiaris*)," *Applied Animal Behaviour Science* 126, nos. 3–4 (2010): 140–53.

104   **initiating more play and using more play faces** Palagi, E., "Social Play in Bonobos (*Pan paniscus*) and Chimpanzees (*Pan troglodytes*): Implications for Natural Social Systems and Interindividual Relationships," *American Journal of Physical Anthropology* 129, no. 3 (2006): 418–26.

and, Wobber, V., R. Wrangham, and B. Hare, "Bonobos Exhibit Delayed Development of Social Behavior and Cognition Relative to Chimpanzees," *Current Biology* 20, no. 3 (2010): 226–30.

104   **Also, feral dogs show more sexual behaviour** see Koler-Matznick, et al., "An Updated Description of the New Guinea Singing Dog."

and, Pal, S., B. Ghosh, and S. Roy. "Inter- and Intra-Sexual Behaviour of Free-Ranging Dogs (*Canis familiaris*)," *Applied Animal Behaviour Science* 62, nos. 2–3 (1999): 267–78.

104   **Their sex lives make chimpanzees and even humans look dull** see De Waal and Lanting, *Bonobo: The Forgotten Ape*.

and, Kano, *The Last Ape*.

and, Savage-Rumbaugh, E. S., and B. J. Wilkerson, "Socio-Sexual Behavior in *Pan paniscus* and *Pan troglodytes*: A Comparative Study," *Journal of Human Evolution* 7, no. 4 (1978): 327–44.

and, Woods, V., and B. Hare, "Bonobo but Not Chimpanzee Infants Use Socio-Sexual Contact with Peers," *Primates* 52, no. 2 (2011): 111–16.

104   **bonobos rarely have been observed hunting** Surbeck, M., et al., "Evidence for the Consumption of Arboreal, Diurnal Primates by Bonobos (*Pan paniscus*)," *American Journal of Primatology* 71, no. 2 (2009): 171–74.

104 **as likely to play with monkeys as they are to try to hunt them** Ihobe, H., "Non-Antagonistic Relations Between Wild Bonobos and Two Species of Guenons," *Primates* 38, no. 4 (1997): 351–57.

104 **have smaller canine teeth** Hare, B., V. Wobber, and R. Wrangham, "The Self-Domestication Hypothesis: Evolution of Bonobo Psychology Is Due to Selection Against Aggression," *Animal Behaviour* 83, no. 3 (2012): 573–85.

104 **chimpanzees lose this when they become juveniles** see Ibid.

105 **canine teeth would improve bonobo survival** see Ibid.

105 **an explanation for how bonobos became so dog-like** see Wrangham and Pilbeam, "African Apes as Time Machines."

106 **a wild species could domesticate themselves** see Zeuner, *A History of Domesticated Animals.*

and, Morey, D. F., "The Early Evolution of the Domestic Dog."

and, Coppinger, R., and L. Coppinger, *Dogs: A New Understanding of Canine Origin, Behavior, and Evolution* (Chicago: University of Chicago Press, 2002).

106 **these have been compared with chimpanzee forests** see Furuichi, "Female Contributions."

and, Malenky, R. K., and R. W. Wrangham, "A Quantitative Comparison of Terrestrial Herbaceous Food Consumption by *Pan paniscus* in the Lomako Forest, Zaire, and *Pan troglodytes* in the Kibale Forest, Uganda," *American Journal of Primatology* 32, no. 1 (1994): 1–12.

and, This is a difficult idea to evaluate because there is so much variability in fruit availability across different bonobo and chimpanzee habitats (see Hohmann, G., et al., "Plant Foods Consumed by Pan: Exploring the Variation of Nutritional Ecology across Africa," *American Journal of Physical Anthropology* 141, no. 3 [2010]: 476–85). In addition, one must assume that current food availability represents past availability. While it is easy to question or debate the fruit productivity hypothesis, there is little room for debate over the absence of gorillas in bonobo forests.

106 **Bonobo females form strong bonds not observed in the less social female chimpanzee** see Wrangham and Peterson, *Demonic Males.*

and, Furuichi, "Female Contributions."

106 **evolutionary advantage to intense male aggression** see Kano, *The Last Ape.*

and, Furuichi, "Female Contributions."

and, Surbeck, Mundry, and Hohmann, "Mothers Matter!"

and, Hare, et al., "The Self-Domestication Hypothesis."

106 **friendly males will be the most successful** Characterizing bonobos as a non-violent ape would be going too far. Bonobo females can severely injure males when they form coalitions against males who become too aggressive. Bonobos are relatively peaceful compared with chimpanzees because aggression has never been observed to lead to death (although see Hohmann, G., and B. Fruth, "Is Blood Thicker Than Water?" In *Among African Apes,* eds. M. Robbins and C. Boesch [Berkeley: University of California at Berkeley Press, 2011] for a potential exception), females are not coerced during mating, and there is no infanticide.

315

107 **such as their smaller teeth and craniums** This also makes the traits that were so difficult to explain, like small teeth and craniums, easily explainable as by-products of this selection against aggression (see Hare, Wobber, and Wrangham, "The Self-Domestication Hypothesis").

108 **dedicated her life to protecting the last bonobos** see André, C., *Une Tendresse Sauvage.*

108 **test our self-domestication hypothesis** We compared the bonobos at Lola ya Bonobo with chimpanzees who are also orphans of the bush-meat trade. Chimpanzees are also threatened by the same problem. While Lola ya Bonobo is the world's only bonobo sanctuary, there are also more than a dozen chimpanzee sanctuaries (see www.pasaprimates.org). We have worked at two of these chimpanzee sanctuaries: the Ngamba Island Chimpanzee Sanctuary and Tchimpounga Chimpanzee Rehabilitation Centre. We were also excited to conduct our research at these sanctuaries because we wanted our efforts to support the sanctuaries while playing a role in protecting bonobos and chimpanzees in the wild.

108 **played together while they were eating** Hare, B., et al., "Tolerance Allows Bonobos to Outperform Chimpanzees on a Cooperative Task," *Current Biology* 17, no. 7 (2007): 619–23.

108 **in some ways bonobos never grow up** We have subsequently found evidence for delayed development of cognitive skills as well (see Wobber, Wrangham, and Hare, "Bonobos Exhibit Delayed Development").

109 **promotes sharing, which supported our prediction** Wobber, V., et al., "Differential Changes in Steroid Hormones before Competition in Bonobos and Chimpanzees," *Proceedings of the National Academy of Sciences* 107, no. 28 (2010): 12457–62.

110 **Bonobos are definitely attracted to strangers** Hare, B., and S. Kwetuenda, "Bonobos Voluntarily Share Their Own Food with Others," *Current Biology* 20, no. 5 (2010): R230–R231.

and, Tan, J., and B. Hare, "Bonobos Share with Strangers," unpublished data.

110 **test from our work with chimpanzees** This ingenious test was originally invented by Satoshi Hirata of Kyoto University (see Hirata, S., and K. Fuwa, "Chimpanzees [*Pan troglodytes*] Learn to Act with Other Individuals in a Cooperative Task," *Primates* 48 [2007]: 13–21).

111 **complexity that is seen in human children** Melis, A. P., B. Hare, and M. Tomasello, "Chimpanzees Recruit the Best Collaborators," *Science* 311, 5765, (2006): 1297–300.

and, Melis, A. P., B. Hare, and M. Tomasello, "Engineering Cooperation in Chimpanzees: Tolerance Constraints on Cooperation," *Animal Behaviour* 72, no. 2 (2006): 275–86.

and, Melis, A. P., B. Hare, and M. Tomasello, "Chimpanzees Coordinate in a Negotiation Game," *Evolution and Human Behavior* 30, no. 6 (2009): 381–92.

111 **making co-operation impossible** The intolerance observed was only a trait of the pairing, not of the individuals. These same individuals were paired with another chimpanzee with whom they would share food. When

re-paired with a tolerant partner, they were able to solve the co-operative problems spontaneously. We also found that individuals who had spontaneously co-operated but were placed with an intolerant partner then failed to co-operate with this new partner. We could turn chimpanzee co-operation on or off just based on their ability to share food with their potential partner (see Melis, Hare, and Tomasello, "Engineering Cooperation in Chimpanzees").

112 **more flexible co-operators than chimpanzees** Hare, B., et al., "Tolerance Allows Bonobos to Outperform Chimpanzees."

and, Dogs and experimental foxes are more skilled at using human social gestures as a result of being self-domesticated. We find a similar difference between bonobos and chimpanzees. Bonobos are more likely to look in the direction a human is looking than are chimpanzees. This means that bonobos are also more sensitive to human social information as a by-product of self-domestication (Herrmann, E., et al., "Differences in the Cognitive Skills of Bonobos and Chimpanzees," *PLoS ONE* 5, no. 8 [2010], e12438). However, while bonobos are more sensitive than chimpanzees, they still fail at the food-finding game where dogs are so successful (Maclean, E., and B. Hare, unpublished data).

112 **self-domestication has made bonobos more intelligent** Hare, B., "From Hominoid to Hominid Mind."

112 **their counterparts living in remaining wild areas** Ditchkoff, S. S., S. T. Saalfeld, and C. J. Gibson, "Animal Behavior in Urban Ecosystems: Modifications Due to Human-Induced Stress," *Urban Ecosystems* 9, no. 1 (2006): 5–12.

112 **living farther away from urban areas** Harveson, P. M., et al., "Impacts of Urbanization on Florida Key Deer Behavior and Population Dynamics," *Biological Conservation* 134, no. 3 (2007): 321–31.

and, Peterson, M., et al., "Wildlife Loss Through Domestication: The Case of Endangered Key Deer," *Conservation Biology* 19, no. 3 (2005): 939–44.

112 **bobcats have begun taking up residence in urban areas** Gehrt, S. D., S. P. D. Riley, and B. L. Cypher, *Urban Carnivores: Ecology, Conflict, and Conservation* (Baltimore: Johns Hopkins University Press, 2010).

113 **learn from adults in ways that other species cannot** see Herrmann, et al., "Humans Have Evolved."

and, Tomasello, et al., "Understanding and Sharing Intentions."

113 **spontaneously help others and enjoy cooperative games** Warneken, F., et al., "Spontaneous Altruism by Chimpanzees and Young Children," *PLOS Biology* 5, no. 7 (2007): e184.

and, Warneken, F., and M. Tomasello, "Varieties of altruism in children and chimpanzees," *Trends in Cognitive Sciences* 13, no. 9 (2009): 397-402.

113 **infants compared with apes** see Herrmann, et al., "Humans Have Evolved."

and, Herrmann, et al., "Differences in the Cognitive Skills."

113 **life has been closely observed across cultures** Hrdy, S. B., *Mothers and Others: The Evolutionary Origins of Mutual Understanding* (Cambridge, Mass.: Belknap Press, 2009).

113 **best of them, and their co-operation fell apart** see Melis, Hare, and Tomasello, "Chimpanzees Recruit."
and, Melis, Hare, and Tomasello, "Engineering Cooperation in Chimpanzees."

113 **we had to become ultra-tolerant** When might have this happened? It all comes down to our last ancestor in common with bonobos and chimpanzees. If this ancestor was more like a bonobo, then our lineage had the benefit of being relatively tolerant from the very start. However, if our ancestor was more like the intolerant chimpanzee, then an even more radical shift in tolerance occurred. Either way, human co-operation could not evolve without a major shift in tolerance. While bonobos are highly tolerant, humans can be even more tolerant towards group members. At least our disputes do not typically end in physical violence.

114 **highly tolerant and socially savvy individuals** When selection against aggression occurred, it would have allowed humans to more peacefully interact in new types of social interactions that would have been impossible before. Any pre-existing variation in cognitive skills relating to these types of social behaviours could then come under direct selection. Nature could favour cleverer, more co-operative humans, since an increase in cognition would pay off in groups of tolerant individuals who could share the spoils of new and more flexible forms of co-operation. Presumably, one of the first cognitive skills that would need to evolve in these cleverer individuals would be flexibility in detecting and avoiding cheaters, so that increases in co-operation could become evolutionarily stable strategies (see Stevens, J. R., F. A. Cushman, and M. D. Hauser, "Evolving the psychological mechanisms for Cooperation," *Annual Review of Ecology, Evolution, and Systematics* 36 [2005]: 499–518; and, Richerson, P. J., and R. Boyd, "The Evolution of Human Ultra-Sociality." In *Indoctrinability, Ideology, and Warfare: Evolutionary Perspectives*, eds. I. Eibl-Eibesfeldt and F. K. Salter [New York: Berghahn Books, 1998], 71–95; and, Barrett, H. C., L. Cosmides, and J. Tooby, "Coevolution of Cooperation, Causal Cognition and Mindreading," *Communicative & Integrative Biology* 3, no. 6 [2010]: 522).

115 **when transporting humans or herding animals** see Coppinger and Schneider, "Evolution of Working Dogs."

116 **more like sight hounds** see Johannes, "Basenji Origin and Migration."

116 **observed in the dogs of the Mayangna people of Nicaragua** Koster, J. M., and K. B. Tankersley, "Heterogeneity of Hunting Ability and Nutritional Status Among Domestic Dogs in Lowland Nicaragua," *Proceedings of the National Academy of Sciences* 109, no. 8 (2012): E463–E470.

116 **largely bred based on their appearance** Scott, J. P., and J. L. Fuller, *Genetics and the Social Behavior of the Dog* (Chicago: University of Chicago Press, 1965).

116 **working dogs are the most skilled of all** We also were able to rule out the alternative explanation for these results, that working dogs are somehow just generally more domesticated. In both our working dog group and our non-working dog group, we had a breed represented that was among the more genetically wolf-like breeds identified (Parker, et al., "Genetic

Structure"). This suggests that heritable variation in the cognitive trait that allows for the use of human gestures can come under direct selection once a dog is self-domesticated.

117 **perhaps most surprisingly – smaller craniums** Leach, H. M., "Human Domestication Reconsidered," *Current Anthropology* 44, no. 3 (2003): 349–68.

and, Hawks, J., "Selection for Smaller Brains in Holocene Human Evolution," arXiv:1102.5604, 2011.

117 **human brain has actually shrunk between 10 and 30 percent** Allman, J., Evolving Brains (New York: Scientific American Library, 1999).

and, Lahr, M. M., *The Evolution of Modern Human Diversity: A Study in Cranial Variation* (Cambridge, UK: Cambridge University Press, 1996).

but, see Kappelman, J., "The Evolution of Body Mass and Relative Brain Size in Fossil Hominids," *Journal of Human Evolution* 30, no. 3 (1996): 243–76.

117 **and China just over 12,000 years ago** Barker, G., *The Agricultural Revolution in Prehistory: Why Did Foragers Become Farmers?* (New York: Oxford University Press, 2009).

118 **during the evolution of our species** see Marlowe, *The Hadza.*

118 **modern populations still living as hunter-gatherers** see Ibid.

and, Hill, K., and A. M. Hurtado, *Ache Life History: The Ecology and Demography of a Foraging People* (New York: Aldine de Gruyter, 1996).

and, Shostak, M., *Nisa: The Life and Words of a !Kung Woman* (Cambridge, Mass.: Harvard University Press, 2000).

and, Ellison, P. T., *On Fertile Ground: A Natural History of Human Reproduction* (Cambridge, Mass.: Harvard University Press, 2003).

118 **even kill anyone who tries to forcibly rule the group** Boehm, C., *Hierarchy in the Forest: The Evolution of Egalitarian Behavior* (Cambridge, Mass.: Harvard University Press, 2001).

118 **lead to more complex technologies** Kline, M. A., and R. Boyd, "Population Size Predicts Technological Complexity in Oceania," *Proceedings of the Royal Society B: Biological Sciences* 277, no. 1693 (2010): 2559–64.

119 **'Humans domesticated dogs, and dogs domesticated humans'** Groves, C., "The Advantages and Disadvantages of Being Domesticated," *Perspectives in Human Biology* 4, no. 1 (1999): 1–12.

120 **Like ravens following wolves on a hunt** see Heinrich, *Mind of the Raven.*

120 **Hunters without dogs are usually less successful** see Koster and Tankersley, "Heterogeneity of Hunting."

120 **more prey when accompanied by dogs** Ruusila, V., and M. Pesonen, "Interspecific Cooperation in Human (*Homo sapiens*) Hunting: The Benefits of a Barking Dog (*Canis familiaris*)," *Annales Zoologici Fennici* 41, no. 4 (2004): 545-49.

120 **also be an emergency food supply** see Clutton-Brock and Hammond, "Hot Dogs."

121 **sacrificed to save the group or the best hunting dogs** see Koster and Tankersley, "Heterogeneity of Hunting."

Chapter 6: Dog Speak (pages 125–45)

126 **he had learned the word so quickly** Oreo was not perfect. Some days when our paper was thrown in the bushes or buried under leaves, Oreo brought the neighbours' paper.

126 **learning words through a process of exclusion** see Kaminski, Call, and Fisher, "Word Learning in a Domestic Dog."

and, Pilley and Reid, "Border Collie Comprehends Object Names."

127 **remember some new names a month later** see Pilley and Reid, "Border Collie Comprehends Object Names."

and, Chaser was tested in a similar way after delays of twenty-four hours and four weeks but was not able to retain any of the new labels without play sessions in which she practised retrieving the new toys at least a few times. Some might suggest that this leaves room for a 'simple' associative account (see Markman, E. M. and M. Abelev, "Word Learning in Dogs?" *Trends in Cognitive Sciences* 8, no. 11 [2004]: 479–81), but I do not see that as being the case, since the initial learning of the pairings occurred as the result of an inference by exclusion. Instead, the initial learning occurs through inference, and the information learned is maintained with a small amount of practice (no rewarding except social praise). This points more to differences in long-term memory in human infants and dogs that allow infants to remember words more skilfully than Chaser (and perhaps Rico). I would be more convinced of a 'simpler' associative account if Chaser learned as many new words as quickly when played on a recording instead of during interactions with Pilley.

127 **how children learn words** A number of apes have undergone intensive training to teach them language skills. Some apes learned to make hand signs using a small number of gestures. Other apes have been able to learn an artificial language called 'Yerkish', named after the founder of American primatology, Robert Yerkes. Both bonobos and chimpanzees can use a lexigram keyboard to communicate with humans. However, their productivity is limited relative to human children. Generally, they can only make requests. I experienced this firsthand when I worked together with a language-trained orangutan named Chantek. He was no longer actively participating in language research at the time, but he still used a few of his signs. He was particularly fond of the signs for 'candy' and 'juice'. A bonobo named Kanzi is the champion of the language-learning experiments. He shows a remarkable capacity for understanding spoken English. This was illustrated in an experiment where a host of novel requests were made in which a large number of words were recombined in ways he had never heard before (see Savage-Rumbaugh, "Language Comprehension in Ape and Child"). When Kanzi was asked to 'put the basketball inside the bowl on the top shelf of the refrigerator', he was able to complete this request (even if he did not understand why such a crazy request was made). However, no one has documented exactly how language-trained apes acquire the skills they demonstrate. The work with Rico and Chaser is unique because it clearly shows the mechanisms by which these dogs are learning the words.

127   **can be different colors, shapes, textures, and sizes** Bloom, P., "Can a Dog Learn a Word?" *Science* 304, no. 5677 (2004): 1605–6.

127   **Chaser's owner, John Pilley** see Pilley and Reid, "Border Collie Comprehends."

128   **Chaser never made a mistake** Chaser's performance discriminating toys from non-toys further diffuses the hypothesis put forward by Markman and Abelev (2004) to explain Rico's word learning. Chaser showed she could spontaneously distinguish between two sets of familiar toys, ruling out the idea that the entire phenomenon is driven by an attraction to novel toys.

129   **She designed a test similar to the one used with children** Mother-reared apes have been tested for their understanding of "iconic" symbols in a very similar set of experiments. They either fail or show limited skills. See Tomasello, M., J. Call, and A. Gluckman, "Comprehension of Novel Communicative Signs by Apes and Human Children," *Child Development* 68, no. 6 (1997): 1067–80.

       and, Herrmann, E., A. Melis, and M. Tomasello, "Apes' Use of Iconic Cues in the Object-Choice Task," *Animal Cognition* 9, no. 2 (2006): 118–30.

129   **other containers of different shapes** Tomasello, Call, Gluckman, "Comprehension of Novel Communicative Signs."

129   **symbolic nature of our helpful behaviour** Kaminski, J., et al., "Domestic Dogs Comprehend Human Communication with Iconic Signs," *Developmental Science* 12, no. 6 (2009): 831–37.

130   **new pictures on a computer screen** see Aust, et al., "Inferential Reasoning by Exclusion."

       and, Erdőhegyi, Á., et al., "Doggy-Computer: Recognizing the Pointing Cue in Two- and Three-Dimension," *Journal of Veterinary Behavior: Clinical Applications and Research* 4, no. 2 (2009): 57.

130   **when an experimenter communicates it** see Kupán, K., et al., "Why Do Dogs (*Canis familiaris*) Select."

130   **hearing a conversation between two people** Following the model-rival method used to train the famous African parrot (Pepperberg, I. M., "Cognitive and Communicative Abilities of Grey Parrots," *Applied Animal Behaviour Science* 100, no. 1–2 [2006]: 77–86), dogs were also trained in spoken labels for a set of novel objects.

131   **more traditional food-rewarding techniques** McKinley, S., and R. J. Young, "The Efficacy of the Model-Rival Method When Compared with Operant Conditioning for Training Domestic Dogs to Perform a Retrieval-Selection Task," *Applied Animal Behaviour Science* 81, no. 4 (2003): 357–65.

131   **barking seems to occur indiscriminately** Coppinger, R., and M. Feinstein, "Hark! Hark! The Dogs Do Bark . . . And Bark . . . And Bark . . . And Bark," *Smithsonian* 21 (1991): 119–29.

132   **the emotional state of the barking dog** Lord, K., M. Feinstein, and R. Coppinger, "Barking and Mobbing," *Behavioural Processes* 81, no. 3 (2009): 358–68.

132 **Barks make up as little as 3 percent of wolf vocalizations** Schassburger, R. M., *Vocal Communication in the Timber Wolf,* Canis lupus, Linnaeus: *Structure, Motivation, and Ontogeny* (Berlin: P. Parey, 1993).

132 **consequence of selecting against aggression** see Gogoleva, et al., "The Sustainable Effect of Selection."

and, Gogoleva, S. S., et al., "Kind Granddaughters of Angry Grandmothers: The Effect of Domestication on Vocalization in Cross-Bred Silver Foxes," *Behavioural Processes* 81, no. 3 (2009): 369–75.

132 **plastic vocal cords, or a 'modifiable vocal tract'** Fitch, W. T., "The Phonetic Potential of Nonhuman Vocal Tracts: Comparative Cineradiographic Observations of Vocalizing Animals," *Phonetica* 57, nos. 2–4 (2000): 205–18.

132 **it turns out that not all barks are the same** Feddersen-Petersen, D. U., "Vocalization of European Wolves (*Canis lupus lupus* L.) and Various Dog Breeds (*Canis lupus* f. fam.)," *Archiv für Tierzucht* 43, no. 4 (2000): 387–98.

132 **barks can vary in timing, pitch, and amplitude** Yin, S., "A New Perspective on Barking in Dogs (*Canis familiaris*)," *Journal of Comparative Psychology* 116, no. 2 (2002): 189–93.

and, Yin, S., and B. McCowan, "Barking in Domestic Dogs: Context Specificity and Individual Identification," *Animal Behaviour* 68, no. 2 (2004): 343–55.

133 **rather than the stranger growl** Faragó, T., et al., "Dogs' Expectation about Signalers' Body Size by Virtue of Their Growls," *PLoS ONE* 5, no. 12 (2010): e15175.

133 **between the barks of different dogs** Maros, K., et al., "Dogs Can Discriminate Barks from Different Situations," *Applied Animal Behaviour Science* 114, no. 1–2 (2008): 159–67.

134 **playing or being aggressive** Pongrácz, P., et al., "Human Listeners Are Able to Classify Dog (*Canis familiaris*) Barks Recorded in Different Situations," *Journal of Comparative Psychology* 119, no. 2 (2005): 136–44.

and, Only the most experienced cat owners can discriminate among different types of cat meows in a similar test setting (Nicastro, N. and M. J. Owren, "Classification of Domestic Cat (*Felis catus*) Vocalizations by Naive and Experienced listeners," *Journal of Comparative Psychology* 117, no. 1 (2003): 44–52.

134 **dogs was when they heard the 'stranger bark'** Molnár, C., et al., "Can Humans Discriminate Between Dogs on the Base of the Acoustic Parameters of Barks?" *Behavioural Processes* 73, no. 1 (2006): 76–83.

135 **Even more amazing, they did this on their very first trial** Kundey, S., et al., "Domesticated Dogs (*Canis familiaris*) React to What Others Can and Cannot Hear," *Applied Animal Behaviour Science* 126, no. 1–2 (2010): 45–50.

136 **they start to understand the gestures of others** Carpenter, et al., "Social Cognition, Joint Attention."

136 **others during their natural interactions** Bekoff, M., "Play Signals as Punctuation: The Structure of Social Play in Canids," *Behaviour* 132, no. 5–6 (1995): 419–29.

136 **that makes them more vulnerable (rolling on their back)** see Ibid.

and, Bauer, E. B., and B. B. Smuts, "Cooperation and Competition During Dyadic Play in Domestic Dogs, *Canis familiaris*," *Animal Behaviour* 73, no. 3 (2007): 489–99.

137 **behaving this way to help me locate the food** see Hare, Call, and Tomasello, "Communication of Food Location."

137 **very persistent at enlisting human help** Gaunet, F., "How Do Guide Dogs and Pet Dogs (*Canis familiaris*) Ask Their Owners for Their Toy and for Playing?" *Animal Cognition* 13, no. 2 (2010): 311–23.

137–8 **a human appears but nothing has been hidden** see Miklósi, Á., et al., "Intentional Behaviour in Dog-Human Communication."

138 **Sofia never used the keyboard when she was alone** Rossi, A. P., and C. Ades, "A Dog at the Keyboard: Using Arbitrary Signs to Communicate Requests," *Animal Cognition* 11, no. 2 (2008): 329–38.

138 **and kept pressing the key for food (as opposed to toy!)** see Ibid.

140 **he nudged the ball into my back and began barking** see Hare, Call, and Tomasello, "Communication of Food Location."

140 **Alexandra Horowitz from Barnard College in New York conducted** Horowitz, A., "Disambiguating the 'Guilty Look': Salient Prompts to a Familiar Dog Behaviour," *Behavioural Processes* 81, no. 3 (2009): 447–52.

140 **who does not respond to visual signals** Gaunet, F., "How Do Guide Dogs of Blind Owners and Pet Dogs of Sighted Owners (*Canis familiaris*) Ask Their Owners for Food?" *Animal Cognition* 11, no. 3 (2008): 475–83.

141 **wearing a blindfold over their mouth** Virányi, Z., et al., "Dogs Respond Appropriately to Cues of Humans' Attentional Focus," *Behavioural Processes* 66, no. 2 (2004): 161–72.

and, Gácsi, M., et al., "Are Readers of Our Face Readers of Our Minds? Dogs (*Canis familiaris*) Show Situation-Dependent Recognition of Human's Attention," *Animal Cognition* 7, no. 3 (2004): 144–53.

and, Fukuzawa, M., D. S. Mills, and J. J. Cooper, "More Than Just a Word: Non-Semantic Command Variables Affect Obedience in the Domestic Dog (*Canis familiaris*)," *Applied Animal Behaviour Science* 91, no. 1–2 (2005): 129–41.

and, Schwab, C., and L. Huber, "Obey or Not Obey? Dogs (*Canis familiaris*) Behave Differently in Response to Attentional States of Their Owners," *Journal of Comparative Psychology* 120, no. 3 (2006): 169–75.

141 **someone wearing a blindfold or sunglasses** One study suggests dogs even know if your eyes are open or shut (see Call, J., et al., "Domestic Dogs [*Canis familiaris*] Are Sensitive to the Attentional State of Humans," *Journal of Comparative Psychology* 117, no. 3 [2003]: 257–63), while another suggests dogs know when you can hear them in some contexts (Kundey, et al., "Domesticated Dogs [*Canis familiaris*] React"). Both are reviewed in Chapter 10 on teaching genius.

142 **though there was no reward for doing so** Kaminski, J., et al., "Domestic Dogs Are Sensitive to a Human's Perspective," *Behaviour* 146, no. 7 (2009): 979–98.

143   **inform adults based on what the adults have seen in the past** Liszkowski, U., et al., "12- and 18-Month-Olds Point to Provide Information for Others," *Journal of Cognition and Development* 7, no. 2 (2006): 173–87.

143   **recognize when someone is knowledgeable or ignorant** see Hare, "From Hominoid to Hominid Mind."

144   **based on what someone had seen** Topál, J., et al., "Reproducing Human Actions and Action Sequences: 'Do as I do!' in a Dog," *Animal Cognition* 9, no. 4 (2006): 355–67.

144   **between people based on what they know or do not know** Virányi, Z., et al., "A Nonverbal Test of Knowledge Attribution: A Comparative Study on Dogs and Children," *Animal Cognition* 9, no. 1 (2006): 13–26.

144   **same method, this finding was not replicated** see Kaminski, "Domestic Dogs Are Sensitive."

144   **understand when others are knowledgeable or ignorant** see Gaunet, "How Do Guide Dogs and Pet Dogs."

      and, Kaminski, J., et al., "Dogs, *Canis familiaris*, Communicate with Humans to Request but Not to Inform." *Animal Behaviour* 82, no. 4 (2011): 651–58.

145   **those observed in a number of other animals** see Cheney and Seyfarth, *Baboon Metaphysics*.

Chapter 7: Lost Dogs (pages 147–64)

148   **they never need to go farther than their front door** Frank, H., "Evolution of Canine Information Processing under Conditions of Natural and Artificial Selection," *Zeitschrift für Tierpsychologie* 53, no. 4 (1980): 389–99.

      and, Fox, M., and D. Stelzner, "Behavioural Effects of Differential Early Experience in the Dog," *Animal Behaviour* 14, no. 2–3 (1966): 273–81.

149   **shortest distance between their location and the reward** see Ibid.

      and, Cattet, J., and A. S. Etienne, "Blindfolded Dogs Relocate a Target Through Path Integration," *Animal Behaviour* 68, no. 1 (2004): 203–12.

      and, Séguinot, V., J. Cattet, and S. Benhamou, "Path Integration in Dogs," *Animal Behaviour* 55, no. 4 (1998): 787–97.

      and, Chapuis, N., and C. Varlet, "Short Cuts by Dogs in Natural Surroundings," *Quarterly Journal of Experimental Psychology* 39, no. 1 (1987): 49–64.

150   **dogs encounter during agility or service dog training** Osthaus, B., D. Marlow, and P. Ducat, "Minding the Gap: Spatial Perseveration Error in Dogs," *Animal Cognition* 13, no. 6 (2010): 881–85.

151   **a big hug from an experimenter** Basenjis performed best in the first trial while beagles performed best overall. Their performances resembled those of the less talented puppies.

      and, Elliot, O., and J. P. Scott, "The Analysis of Breed Differences in Maze Performance in Dogs," *Animal Behaviour* 13, no. 1 (1965): 5–18.

151   **until their performance resembled the less talented puppies'** see Ibid.

153   **dogs were correct only 55 percent of the time** Another study (Macpherson, K., and W. A. Roberts, "Spatial Memory in Dogs [*Canis familiaris*] on a

Radial Maze," *Journal of Comparative Psychology* 124, no. 1 [2010]: 47) used the same eight-armed maze to test the ability of dogs to remember the location of four sites they had previously visited. The dogs were prevented from visiting four of the arms and allowed to visit the remaining arms. After a delay, they were allowed back into the maze, which still contained food in the arms they had not visited. The rats outperformed the dogs by an even greater margin.

To date, only Rico has shown an ability to remember what has been hidden in a way that is comparable to what has been observed in other species such as rats and birds (see Kamil, A. C., R. P. Balda, and D. J. Olson, "Performance of Four Seed-Caching Corvid Species in the Radial-Arm Maze Analog," *Journal of Comparative Psychology* 108, no. 4 [1994]: 385; and, Bird, L. R., et al., "Spatial Memory for Food Hidden by Rats [*Rattus norvegicus*] on the Radial Maze: Studies of Memory for Where, What, and When," *Journal of Comparative Psychology* 117, no. 2 [2003]: 176). Rico was almost flawless when he had to remember in which room different toys had been placed when he was asked to retrieve them (see Kaminski, J., J. Fischer, and J. Call, "Prospective Object Search in Dogs: Mixed Evidence for Knowledge of What and Where," *Animal Cognition* 11, no. 2 [2008]: 367–71).

153 **though his owner had moved four times** Johnston, L., "Missing dog finds way home after 5 years – even to owners' new house," *New York Daily News*, January 19, 2011, http://articles.nydailynews.com/2011-01-19/entertainment/27088069_1_missing-dog-children-prince.

153 **and came home despite two broken legs** West, K., "Mason the 'Tornado Dog' Finds His Way Home on Two Legs," *People*, June 13, 2011; http://www.peoplepets.com/people/pets/article/0,20502339,00.html.

153 **find a target from multiple positions** As humans, we can use different strategies to explain directions to someone. You can either explain where a location is relative to your current position by indicating a direction (say, left or right) or by explaining the target's location relative to a well-known landmark. These two strategies are similar to different strategies observed in animals as they navigate. Many animals are able to use their own location in space to remember the position of other objects (*the nut is to my left*); some species can even use landmarks to navigate (*the nut is next to the tree*).

154 **search proportionally to the new position of the peg** Fiset, S., "Landmark-Based Search Memory in the Domestic Dog (*Canis familiaris*)," *Journal of Comparative Psychology* 121, no. 4 (2007): 345–53.

and, Milgram, N. W., et al., "Landmark Discrimination Learning in the Dog: Effects of Age, an Antioxidant Fortified Food, and Cognitive Strategy," *Neuroscience & Biobehavioral Reviews* 26, no. 6 (2002): 679–95.

and, Milgram, N. W., et al., "Landmark Discrimination Learning in the Dog," *Learning & Memory* 6, no. 1 (1999): 54–61.

154 **This is called using an egocentric approach** Fiset, S., S. Gagnon, and C. Beaulieu, "Spatial Encoding of Hidden Objects in Dogs (*Canis familiaris*)," *Journal of Comparative Psychology* 114, no. 4 (2000): 315–24.

154 **"Most lost dogs never find their homes"** Miklósi, *Dog Behavior, Evolution and Cognition.*

154 **show an understanding of basic physical principles** see Herrmann, et al., "Humans Have Evolved Specialized."

and, Spelke, E. S., et al., "Origins of Knowledge," *Psychological Review* 99, no. 4 (1992): 605.

154 **toys fall down if they are dropped** Hood, B. M., L. Santos, and S. Fieselman, "Two-Year-Olds' Naive Predictions for Horizontal Trajectories," *Developmental Science* 3, no. 3 (2000): 328–32.

154 **pass through a solid object like a wall** Hood, B., S. Carey, and S. Prasada, "Predicting the Outcomes of Physical Events: Two-Year-Olds Fail to Reveal Knowledge of Solidity and Support," *Child Development* 71, no. 6 (2000): 1540–54.

154 **you move one, the other will move, too** see Herrmann, et al., "Humans Have Evolved Specialized."

155 **dogs never solved the more complicated versions** Frank, H., and M. G. Frank, "Comparative Manipulation-Test Performance in Ten-Week-Old Wolves (*Canis lupus*) and Alaskan Malamutes (*Canis familiaris*): A Piagetian Interpretation," *Journal of Comparative Psychology* 99, no. 3 (1985): 266–74.

155 **pull food out of a transparent box** Osthaus, B., S. E. G. Lea, and A. M. Slater, "Dogs (*Canis lupus familiaris*) Fail to Show Understanding of Means-End Connections in a String-Pulling Task," *Animal Cognition* 8, no. 1 (2005): 37–47.

157 **They did not understand that the string needed to be connected to the food** see Ibid.

157 **skilled at solving a series of related problems** see Herrmann, et al., "The Structure of Individual Differences."

and, Herrmann, et al., "Differences in the Cognitive Skills."

and, Heinrich, B., and T. Bugnyar, "Testing Problem Solving in Ravens: String-Pulling to Reach Food," *Ethology* 111, no. 10 (2005): 962–76.

and, Santos, L. R., et al., "Means-Means-End Tool Choice in Cotton-Top Tamarins (*Saguinus oedipus*): Finding the Limits on Primates' Knowledge of Tools," *Animal Cognition* 8, no. 4 (2005): 236–46.

157 **although some dogs may succeed on related tasks** Range, F., M. Hentrup, and Z. Virányi, "Dogs Are Able to Solve a Means-End Task," *Animal Cognition* 14, no. 4 (2011): 575–83.

157 **get tangled up on the same tests** Whitt, E., et al., "Domestic cats (Felis catus) do not show causal understanding in a string-pulling task," *Animal Cognition* 12, no. 5 (2009): 739–43.

158 **made a noise when it was shaken or not** see Bräuer, et al., "Making Inferences About."

158 **for choosing the board on an incline** A control ruled out the possibility that dogs were simply attracted to boards displayed at an incline. They did not always show a preference for an inclined board. When they were rewarded for choosing the inclined board in trials where they knew a piece of wood was hidden instead of food, they did not choose the inclined board. This also

makes it unlikely that the positive result in the experimental condition was due to experimental cues given by touching the boards, since the baiting procedures were identical in the two conditions.

and, Bräuer, et al., "Making Inferences About."

159 **pass through it into the more distant part of the box** Kundey, S. M. A., et al., "Domesticated Dogs' (*Canis familiaris*) Use of the Solidity Principle," *Animal Cognition* 13, no. 3 (2010): 497–505.

160 **improve their performance with practice** Osthaus, B., A. M. Slater, and S. E. G. Lea, "Can Dogs Defy Gravity? A Comparison with the Human Infant and a Non-Human Primate," *Developmental Science* 6, no. 5 (2003): 489–97.

160 **dropping food to defy gravity** This experiment has been critiqued on the fact that the dogs may have used the human's hand movements above the initial drop point in making their decision. In addition, to show a gravity bias it is important to show that when the apparatus is laid flat, the dogs can predict where the food will go through the vertical tube (see Hood, Santos, and Fieselman, "Two-Year-Olds' Naive Predictions").

160 **surprise if a different type of food is removed** Bräuer, J., and J. Call, "The Magic Cup: Great Apes and Domestic Dogs (*Canis familiaris*) Individuate Objects According to Their Properties," *Journal of Comparative Psychology* 125, no. 3 (2011): 353–61.

160–61 **to help them when they encounter an impossible problem** Míklósi, Á., et al., "A Simple Reason for a Big Difference: Wolves Do Not Look ack at Humans, But Dogs Do," *Current Biology* 13, no. 9 (2003): 763–66.

161 **lean over to inspect inside the tubes to find the food** Call, J., and M. Carpenter, "Do Apes and Children Know What They Have Seen?" *Animal Cognition* 3, no. 4 (2001): 207–20.

161 **even though they did not know where the food was** Bräuer, J., J. Call, and M. Tomasello, "Visual Perspective Taking in Dogs (*Canis familiaris*) in the Presence of Barriers," *Applied Animal Behaviour Science* 88, no. 3–4 (2004): 299–317.

and, McMahon, S., K. Macpherson, and W. A. Roberts, "Dogs Choose a Human Informant: Metacognition in Canines," *Behavioural Processes* 85, no. 3 (2010): 293–98.

161 **to inspect both hiding locations** see Udell, Giglio, and Wynne, "Domestic Dogs (*Canis familiaris*) Use Human Gestures."

and, It is possible that dogs implicitly know they do not know but are too impatient to show the type of inspection behaviour used as the measure of understanding in these two studies (that is, they likely discount future payoffs too heavily). Future studies requiring less patience or explicit choices might reveal an implicit understanding of self-knowledge in dogs.

162 **no sign of understanding their mirror image** see Tomasello and Call, *Primate Cognition*.

and, Zazzo, R., "Des Enfants, Des Singes et Des Chiens Devant le Miroir," *Revue de Psychologie Appliquée* 29, no. 2 (1979): 235–46.

and, Howell, T. J., and P. C. Bennett, "Can Dogs (*Canis familiaris*) Use a Mirror to Solve a Problem?" *Journal of Veterinary Behavior: Clinical Applications and Research* 6, no. 6 (2011): 306–12.

162 **discriminate his advertisements from other dogs'** Bekoff, M., "Observations of Scent-Marking and Discriminating Self from Others by a Domestic Dog (*Canis familiaris*): Tales of Displaced Yellow Snow," *Behavioural Processes* 55, no. 2 (2001): 75–79.

163 **reflect upon what they know and do not know** Given how new this area of research is, much future research will be needed to know for sure how well dogs understand themselves. This represents one of the most challenging areas of animal research. I suspect this could be an area where dogs will surprise us in the future. This is largely because making inferences through exclusion may need some rudimentary form of self-knowledge or meta-cognition. (Perhaps Rico can only pair an unfamiliar noise with an unfamiliar toy spontaneously because he is aware which toys he knows or does not know the labels for.)

163 **learning to associate a colored block with the location of food** see Frank, "Evolution of Canine Information Processing."

163 **longer than the wolves to master the task** see Frank, et al., "Motivation and Insight in Wolf."
   and, Frank, H., "Wolves, Dogs, Rearing and Reinforcement: Complex Interactions Underlying Species Differences in Training and Problem-Solving Performance," *Behavior Genetics* 41, no. 6 (2011): 830–39.

163 **dogs should not go up against a lone wolf** Frank had previously compared the malamute puppies' performances to those of wolves who were reared by their mothers and had little contact with humans. These wolves actually did not outperform the dogs. Frank attributes the different performances in the wolves to motivational differences depending on their rearing history. Apparently he observed that the mother-reared wolf puppies were less food-motivated than hand-reared wolf puppies, whom he reported as having a voracious appetite (see Frank, et al., "Motivation and Insight in Wolf").

164 **while dogs took much longer** Wobber, V., and B. Hare, "Testing the Social Dog Hypothesis: Are Dogs Also More Skilled Than Chimpanzees in Non-Communicative Social Tasks?" *Behavioural Processes* 81, no. 3 (2009): 423–28.

Chapter 8: Pack Animals (pages 165–87)

166 **attracted to odours of new foods from the mouths of healthy rats** Kuan, L., and R. M. Colwill, "Demonstration of a Socially Transmitted Taste Aversion in the Rat," *Psychonomic Bulletin & Review* 4, no. 3 (1997): 374–77.

166 **they tended to prefer the same flavour** Lupfer-Johnson, G., and J. Ross, "Dogs Acquire Food Preferences from Interacting with Recently Fed Conspecifics," *Behavioural Processes* 74, no. 1 (2007): 104–6.

167 **other dogs eating than when eating alone** Ross, S., and J. G. Ross, "Social Facilitation of Feeding Behavior in Dogs: I. Group and Solitary Feeding," *The Pedagogical Seminary and Journal of Genetic Psychology* 74, no. 1 (1949): 97–108.

167 **food around barriers faster and with fewer errors than dogs** see Frank, et al., "Motivation and Insight in Wolf."

167 **detour around a V-shaped fence** Pongrácz, P., et al., "Social Learning in Dogs: The Effect of a Human Demonstrator on the Performance of Dogs in a Detour Task," *Animal Behaviour* 62, no. 6 (2001): 1109–17.

and, Pongrácz, P., et al., "Interaction Between Individual Experience and Social Learning in Dogs," *Animal Behaviour* 65, no. 3 (2003): 595–603.

and, Pongrácz, P., et al., "Preference for Copying Unambiguous Demonstrations in Dogs (*Canis familiaris*)," *Journal of Comparative Psychology* 117, no. 3 (2003): 337–43.

and, Pongrácz, P., et al., "Verbal Attention Getting as a Key Factor in Social Learning Between Dog (*Canis familiaris*) and Human," *Journal of Comparative Psychology* 118, no. 4 (2004): 375–83.

168 **dogs moved the door in the same direction they saw demonstrated by another dog** see Miller, Rayburn-Reeves, and Zentall, "Imitation and Emulation by Dogs." Eleven of twelve dogs matched the direction that the demonstrator dog used on the first trial, which is highly significant. While nine of twelve dogs tested with a human demonstrator also matched the direction demonstrated, this was not statistically above chance. It may be that dogs are slightly better at copying dogs, but future research will be needed that is designed to test this possibility.

168 **dogs did not spontaneously copy a human** Kubinyi, E., et al., "Dogs (*Canis familiaris*) Learn from Their Owners Via Observation in a Manipulation Task," *Journal of Comparative Psychology* 117, no. 2 (2003): 156–65.

169 **move the handle in the same direction** In one study, while dogs quickly learned through observation to move a handle to release a toy, they did not copy the direction of movement they observed during the demonstration (see Ibid.; and, Mersmann, D., et al., "Simple Mechanisms Can Explain Social Learning in domestic dogs [*Canis familiaris*]," *Ethology* 117, no. 8 [2011]: 675–90). However, there are mixed results showing that dogs do not always follow even the simplest of target actions (i.e., the direction of movement). Based on this, it seems unlikely that dogs are able to socially learn to reproduce a more complex series of novel actions that would be required to solve even the simplest problems that we face daily.

169 **teach them to do as you say and not as you do** Range, F., L. Huber, and C. Heyes, "Automatic Imitation in Dogs," *Proceedings of the Royal Society B: Biological Sciences* 278, no. 1703 (2011): 211–17.

169 **service dog we have already met named Phillip** see Topál, et al., "Reproducing Human Actions."

170 **between the different positions the same way Topál had** see Ibid.

170 **watched an adult turn on a lamp with their head** Gergely, G., H. Bekkering, and I. Király, "Rational Imitation in Preverbal Infants," *Nature* 415, no. 6873 (2002): 755.

170 **used their hands to turn on the lamp** In this way developmental psychologists have been able to show that young infants not only understand the behaviour of others as intentional but they also understand that others can choose between different actions to accomplish their goal (e.g.,

depending on constraints in the environment). This understanding gives children tremendous flexibility in learning from others as they grow.

172 **analogous to the finding with infants** Range, F., Z. Viranyi, and L. Huber, "Selective Imitation in Domestic Dogs," *Current Biology* 17, no. 10 (2007): 868–72.

172 **to replicate the results** When researchers make a remarkable finding, it is important that other groups replicate their findings. This finding is now in question since a new study replicated the exact original method (see Range, et al., "Selective Imitation") but did not find any evidence of rational imitation (Kaminski, J., et al., "Do Dogs Distinguish Rational from Irrational Acts?" *Animal Behaviour* 81, no. 1 [2011]: 195–203). A much larger sample of dogs participated but did not copy the method of the doggy demonstrator based on whether they had a ball in their mouth or not. Kaminski and colleagues (Kaminski, above) also failed to replicate a related finding (Hauser, et al., "What Experimental Experience") that utilized a different method, suggesting dogs might infer the rationale behind others' behaviour.

172 **dependent on pack power** Further studies showed that there were limits to the dogs' social learning in this detour test. While they learned after one demonstration how to round the detour, they copied the demonstrators' direction only if the demonstrators returned from behind the barrier on the same side from which they originally went behind the barrier. Also see Mersmann, et al., ("Simple Mechanisms"), who replicated the procedure but found slightly different results that argue for a very simple social learning mechanism.

172 **dingoes and New Guinea Singing Dogs** Because dingoes have been observed to live completely independently from humans in the Australian Outback, some scientists believe they should not be considered feral dogs (see Koler-Matznick, "The Origin of the Dog Revisited"). We consider them feral dogs here because genetic studies have shown that dingoes are domestic dogs. Both dingoes and New Guinea Singing Dogs are likely highly similar to the domestic dogs who first began living with humans thousands of years ago. They may have never adopted a lifestyle of full dependence on humans (see Spotte, *Societies of Wolves*).

172 **stray dogs who survive by scavenging human refuse** see Ibid.

172 **intentionally bred by humans for generations** Boitani, L., P. Ciucci, and A. Ortolani, "Behaviour and Social Ecology of Free-Ranging Dogs." In *The Behavioural Biology of Dogs*, ed. P. Jensen (Wallingford, UK: CAB International, 2007), 147–65.

172–3 **few dogs but can reach a stable size of more than ten** see Daniels and Bekoff, "Feralization."

and, Boitani and Ciucci, "Comparative Social Ecology."

and, Purcell, *Dingo.*

and, Bonanni, et al., "Free-Ranging Dogs."

and, Pal, S., B. Ghosh, and S. Roy, "Dispersal Behaviour of Free-Ranging Dogs (*Canis familiaris*) in Relation to Age, Sex, Season and Dispersal Distance," *Applied Animal Behaviour Science* 61, no. 2 (1998): 123–32.

and, Cafazzo, S., et al., "Dominance in Relation to Age, Sex, and Competitive Contexts in a Group of Free-Ranging Domestic Dogs," *Behavioral Ecology* 21, no. 3 (2010): 443–55.

173 **dominant to subadults and juveniles** Not all researchers of feral dogs agree there is a clear hierarchy in feral dog groups. In some groups, a hierarchy has not been observed (see Boitani, Ciucci, and Ortolani, "Behaviour and Social Ecology"; and, Bradshaw, J. W. S., E. J. Blackwell, and R. A. Casey, "Dominance in Domestic Dogs – Useful Construct or Bad Habit?" *Journal of Veterinary Behavior: Clinical Applications and Research* 4, no. 3 [2009]: 135–44). However, in a recent carefully designed study, a hierarchy was detected (see Cafazzo, et al., "Dominance in Relation to Age"). In this group, younger dogs showed submissive behaviours towards adults, while males were typically (but not always) dominant to females in each age category. Submissive behaviours, as opposed to dominance signals, are thought to be the most reliable indicators of dominance hierarchies in mammals such as carnivores and primates. Wolves have a formal dominance signal to display subordination. This means this signal is used only to indicate subordination (it is never used to initiate play or the like). Younger individuals bow under older individuals and lick their muzzles rapidly. It is typically used in affiliative contexts by wolves during greetings or when a group re-assembles after being separated. Feral dogs also show the same formal signal of subordination to one another, but it is not displayed by all individuals and occurs at very low frequency. This low frequency suggests that the dominance hierarchy observed in some feral dog packs is not as absolute as that observed in wolf packs, since dogs seem more relaxed about indicating their subordination. It also is tempting to speculate whether dogs are licking our faces as a sign of affiliation or to indicate subordination. We may need a clever experiment to find out.

173 **a positive interaction shortly after a conflict is a powerful mechanism** Aureli, F., *Natural Conflict Resolution* (Berkeley: University of California Press, 2000).

173 **feral dogs approach cooperation differently from wolves** Dingoes might be the exception, since they are suspected to have very wolf-like social structure in some cases (see Purcell, *Dingo*).

173 **a single breeding pair is dominant to everyone else** see Derix, et al., "Male and Female Mating Competition."

173 **the breeding of other pack members** In its extreme forms, female suppression can take the form of infanticide, where the breeding pair will kill any pups born to another in the group (see Ibid.; and, McLeod, "Infanticide by Female Wolves"). Exceptions to this rule have been observed when mothers tolerate their daughters' offspring (Mech and Boitani, *Wolves*).

173 **attacks to prevent other females from mating** see Derix, et al., "Male and Female Mating Competition."

and, McLeod, "Infanticide by Female Wolves."

and, McLeod, P. J., et al., "The Relation Between Urinary Cortisol Levels and Social Behaviour in Captive Timber Wolves," *Canadian Journal of Zoology* 74, no. 2 (1996): 209–16.

and, Sands, J., and S. Creel, "Social Dominance, Aggression and Faecal Glucocorticoid Levels in a Wild Population of Wolves, *Canis lupus*," *Animal Behaviour* 67, no. 3 (2004): 387–96.

173 **the dominant female will often kill her pups** see McLeod, "Infanticide by Female Wolves."

174 **This may be something similar to jealousy** see Ibid.

174 **they are fully grown would be dangerous** see Mech and Boitani, *Wolves*.

174 **while their parents are hunting** Hunter-gatherers are known as central-place foragers because everyone comes back to camp each day and shares their harvest (see Marlowe, *The Hadza*). Wolves also return each day to a central den to share food with the youngest members of the pack – typically the year's pups (see Mech and Boitani, *Wolves*). In this way humans are more similar to wolves than they are to other primates. This trait in wolves likely made it easy for proto-dogs to replace the den with life near or in a human settlement.

174 **increasing the chance that their newest offspring will survive as well** see Ibid.

174 **this hierarchy is not as strict as in wolves** see Cafazzo, et al., "Dominance in Relation."

174 **There is no dominant pair that leads the group** see Ibid.
and, Boitani and Ciucci, "Comparative Social Ecology."
and, Pal, "Parental Care in Free-Ranging Dogs."

174 **the dog who has the most friends** Bonanni, R., et al., "Effect of Affiliative and Agonistic Relationships on Leadership Behaviour in Free-Ranging Dogs," *Animal Behaviour* 79, no. 5 (2010): 981–91.

174 **no single breeding pair** The possible exceptions to this are the cases of the New Guinea Singing Dog and the dingo. It is possible that some co-operative rearing of offspring occurs in these populations, although the necessary studies have yet to be conducted (see Purcell, *Dingo*).

174 **Instead, dogs are promiscuous** see Boitani and Ciucci, "Comparative Social Ecology."
and, Pal, S. K., "Reproductive Behaviour of Free-Ranging Rural Dogs in West Bengal, India," *Acta Theriologica* 48, no. 2 (2003): 271–81.

174 **prevent subordinate females from breeding** One group did report observing several instances of large male dogs 'forcing themselves' on young females who had just reached reproductive age (see Ghosh, B., D. K. Choudhuri, and B. Pal, "Some Aspects of the Sexual Behaviour of Stray Dogs, *Canis familiaris*," *Applied Animal Behaviour Science* 13, no. 1 [1984]: 113–27). While this is coercive behaviour, it is not meant to prevent the females from having offspring – quite the reverse. So while not admirable, it is not considered reproductive suppression.

175 **little help rearing their puppies** Mech, et al., *The Wolves of Denali*.
and, Spotte, *Societies of Wolves*.

175 **high mortality in feral dog puppies** see Pal, "Maturation and Development."
and, Boitani and Ciucci, "Comparative Social Ecology."

and, Exceptions have been observed. In one case, a pair of feral dog mothers were observed co-nursing each other's puppies. In another case, a male feral dog provided regurgitated food to his suspected litter for ten days while the mother was away foraging (see Pal, "Parental Care in Free-Ranging Dogs").

175 **overthrow their father as the dominant male in the group** Jenks, S. M., *Behavioral Regulation of Social Organization and Mating in a Captive Wolf Pack* (Storrs, Conn.: University of Connecticut, 1988).

175 **coalitions are difficult to observe in feral dogs** No observational study of dogs has reported coalitionary behaviour of adults. This is probably because aggression in feral dogs is extremely rare. Zero cases of aggression were observed in a long-term study of more than two dozen feral dogs when no food or oestrous females were present to fight over (see Cafazzo, et al., "Dominance in Relation to Age"). Coalitionary behaviour in wolves is over disputes for access to the breeding female. Dogs are promiscuous and have multiple breeding females in a group, which likely decreases incentive for coalitionary aggression as well.

176 **will almost always play-attack the dog who is losing** see Bauer and Smuts, "Cooperation and Competition."
and, Ward, C., R. Trisko, and B. B. Smuts, "Third-Party Interventions in Dyadic Play Between Littermates of Domestic Dogs, *Canis lupus familiaris*," *Animal Behaviour* 78, no. 5 (2009): 1153–60.

176 **puppies jointly attack others while playing** Why adult feral dogs do not show coalitionary support is an interesting area for future research. It is possible that further observation will show a low rate of coalitionary support during conflicts or that in some contexts there is a high degree of such support. Alternatively, it might be that adult feral dogs simply do not express this type of co-operative behaviour – perhaps due to the changes in their emotional reactivity or social organization. Based on my own anecdotal observations of pet dogs at off-lead parks, a careful observational study might reveal the presence of coalitionary support during spontaneous dogfights. It is also possible that dogs are more likely to attack other dogs when their owner is in close proximity because they expect support from their human partner. There may also be interesting breed differences. This is clearly another interesting area for future research.

176 **attack and kill members of their own species** Wrangham, R. W., "Evolution of Coalitionary Killing," *American Journal of Physical Anthropology* supplement 29 (1999): 1–30.

176 **detects a lone wolf, they will try to catch and kill her** Murray, D. L., et al., "Death from Anthropogenic Causes Is Partially Compensatory in Recovering Wolf Populations," *Biological Conservation* 143, no. 11 (2010): 2514–24.

176 **of wolves were killed by other wolves** see Mech, et al., *The Wolves of Denali*.
and, Murray, et al., "Death from Anthropogenic."
and, Mech, L. D., "Buffer Zones of Territories of Gray Wolves as Regions of Intraspecific Strife," *Journal of Mammalogy* 75, no. 1 (1994): 199–202.

176 **no one has ever reported them killing one another** Although one suspected case has been observed (see Macdonald and Carr, "Variation in Dog Society").

176 **they tend to have a barking competition** see Bonanni, et al., "Free-Ranging Dogs."

176 **rare when compared to wolves** Several research groups have observed the aggressive behaviour between different feral dog packs (see Boitani and Ciucci, "Comparative Social Ecology"; Pal, Ghosh, and Roy, "Agonistic Behaviour"; Bonanni and Ciucci, "Free-Ranging Dogs"; Bonanni, R., P. Valsecchi, and E. Natoli, "Pattern of Individual Participation and Cheating in Conflicts Between Groups of Free-Ranging Dogs," *Animal Behaviour* 79, no. 4 [2010], 957–68). Bonanni and colleagues (Bonanni and Ciucci, "Free-Ranging Dogs"; and, Bonanni, Valsecchi, and Natoli, "Pattern of Individual Participation") observed a physical fight that included biting in less than 5 percent of the almost two hundred pack disputes witnessed.

176 **crucial strategy for outcompeting other packs** see Bonanni and Ciucci, "Free-Ranging Dogs."

177 **one another, in a similar way to lions** see Mech, *The Wolf.*

and, Stander, P., "Cooperative Hunting in Lions: The Role of the Individual," *Behavioral Ecology and Sociobiology* 29, no. 6 (1992): 445–54.

and, Although wolves likely co-ordinate their behaviour during hunts, no experiment has ever been conducted to test this idea. Both hyenas and chimpanzees are also suspected of co-ordinating their behaviour during group hunts in the wild. Experiments have shown they can solve a novel food-acquisition game by co-ordinating their efforts (Melis, Hare, and Tomasello, "Chimpanzees Recruit the Best"; Melis, Hare, and Tomasello, "Chimpanzees Coordinate"; Melis, A. P., B. Hare, and M. Tomasello, "Do Chimpanzees Reciprocate Received Favours?" *Animal Behaviour* 76, no. 3 [2008]: 951–62.; Drea, C. M. and A. N. Carter, "Cooperative Problem Solving in a Social Carnivore," *Animal Behaviour* 78, no. 4 [2009]: 967–77). Similar experiments will be needed to understand whether wolves are simply acting together when they hunt or if they are actively coordinating their attack.

177 **feral dogs are poor hunters** see Boitani and Ciucci, "Comparative Social Ecology."

177 **sources of food such as rubbish dumps** see Coppinger and Coppinger, *Dogs: A New Understanding.*

and, Spotte, *Societies of Wolves.*

The dingo (and the New Guinea Singing Dog) is an important exception to this rule. Dingoes are best described as being hypercarnivorous, since "they live in groups, eat more than 70% vertebrate prey and hunt animals weighing more than their average weight" (see Purcell, *Dingo*). Arguably, dingoes do not rely on humans and by some accounts should not be considered feral dogs at all. Some have suggested they fill the ecological niche of the North American coyote in Australia (see Spotte, *Societies of Wolves*).

177 **When feral dogs hunt, they are not usually successful** see Boitani and Ciucci, "Comparative Social Ecology."

177 **The exception is dogs who prey on species that did not evolve with large mammalian predators** see Purcell, *Dingo*.
   and, Kruuk, H., and H. Snell, "Prey Selection by Feral Dogs from a Population of Marine Iguanas (*Amblyrhynchus cristatus*)," *Journal of Applied Ecology* 18 (1981): 197–204.

177 **like the co-ordination seen in other mammals** see Mech, *The Wolf*.
   and, Lwanga, et al., "Primate Population Dynamics."
   and, Stander, "Cooperative hunting in Lions."
   and, Drea and Carter, "Cooperative Problem Solving."
   but, A possible exception might be the capture of wallabies and kangaroos by dingoes. While they typically make up a small portion of dingoes' diet, these species can be larger than dingoes as adults and are skilled at escape. This may suggest that co-ordination among multiple individuals is necessary in capturing them. However, no systematic study has been carried out to examine how dingoes catch these larger prey species (see Purcell, *Dingo*).

177 **people to detect and occasionally capture prey** Chowdhury, B., et al., *Behavioral Genetic Characterization of Hunting in Domestic Dogs*, Canis familiaris (Bowling Green, Ohio: Bowling Green State University, 2011).

177 **The hunters then shoot the prey with a dart or arrow** see Koster, "Hunting with Dogs in Nicaragua."
   and, Koster, "The Impact of Hunting with Dogs."

179 **Darwin wrote that his "savage" dog, Czar, did not growl** see Townshend, *Darwin's Dogs*.

180 **separation unless they had been living with them** Hepper, P. G., "Long-Term Retention of Kinship Recognition Established During Infancy in the Domestic Dog," *Behavioural Processes* 33, nos. 1–2 (1994): 3–14.

180 **they also remember what their owner looks like** Adachi, I., H. Kuwahata, and K. Fujita, "Dogs Recall Their Owner's Face Upon Hearing the Owner's Voice," *Animal Cognition* 10, no. 1 (2007). 17–21.

181 **the best co-operative partners** Kundey, S. M. A., et al., "Reputation-like Inference in Domestic Dogs (*Canis familiaris*)," *Animal Cognition* 14, no. 2 (2010): 291–302.

181 **disputes to minimize the risk of injuries** see Bonanni, Valsecchi, and Natoli, "Pattern of Individual Participation."

181 **solve the problem on their own** Miklósi, et al., "A Simple Reason for a Big Difference."

181 **towards items they need help in obtaining** see Hare, Call, and Tomasello, "Communication of Food Location."
   and, Miklósi, et al., "Intentional Behaviour."
   and, Gaunet, "How Do Guide Dogs and Pet Dogs."

181 **dogs were not very good at doing so** Horn, L., et al., "Domestic Dogs (*Canis familiaris*) Flexibly Adjust Their Human-Directed Behavior to the Actions of Their Human Partners in a Problem Situation," *Animal Cognition* 15, no. 1 (2011): 57–71.

182 **five-month-old infants have basic counting skills** Wynn, K., "Addition and Subtraction by Human Infants," *Nature* 358, no. 6389 (1992): 749–50.

182 **number they remembered the experimenter hiding** West, R. E., and R. J. Young, "Do Domestic Dogs Show Any Evidence of Being Able to Count?" *Animal Cognition* 5, no. 3 (2002): 183–86.

182 **to choose between three and two pieces** Dogs could discriminate between 1 versus 4; 1 versus 3; 2 versus 5; 1 versus 2; 2 versus 4; 3 versus 5, but they failed to discriminate between 2 versus 3 or 3 versus 4. Their performance obeyed Weber's law in that dogs were more likely to correctly choose larger quantities when the ratio of the two quantities was small (ratios calculated by dividing smaller quantity by larger) and when the numerical distance between the two numbers was larger (see Ward, C., and B. B. Smuts, "Quantity-Based Judgments in the Domestic Dog [*Canis lupus familiaris*]," *Animal Cognition* 10, no. 1 [2007]: 71–80).

182 **observed in Italian feral dogs** It should be noted that in all of these studies, group statistics are used. When we have designed tests to examine the ability of individual dogs to discriminate quantities as larger or smaller, we rarely see performances that are above chance at the individual level. This suggests that dogs have a very weak ability to discriminate quantity in these settings.

182 **bigger pack who went on the offensive** see Bonanni, et al., "Free-Ranging Dogs."

183 **odds are in their favour based on the relative size of the opposing pack** see Ibid.

183 **believe their dog feels guilty for disobeying** Morris, P. H., C. Doe, and E. Godsell, "Secondary Emotions in Non-Primate Species? Behavioural Reports and Subjective Claims by Animal Owners," *Cognition and Emotion* 22, no. 1 (2008): 3–20.

184 **does not look like the dogs were feeling guilty** see Horowitz, "Disambiguating the 'Guilty Look.'"

184 **emotional responses to co-operative outcomes** Testing whether primates have a sense of fairness has been one of the most controversial areas in the field of animal psychology in the past decade. While there is evidence that can be interpreted in favour of a sense of fairness in primates, the majority of studies argue against a sense of fairness (although see De Waal, F. B. M., "Putting the Altruism Back into Altruism: The Evolution of Empathy," *Annual Review of Psychology* 59 [2008]: 279–300). While there is a single study consistent with a sense of fairness in dogs, it should be viewed with caution, given the mixed findings with primates. This study of fairness in dogs (see Range, F., et al., "The Absence of Reward Induces Inequity Aversion in Dogs," *Proceedings of the National Academy of Sciences* 106, no. 1 [2009]: 340–46) will need to be replicated by other labs before it is widely accepted by most animal psychologists. To my knowledge, no replications had been published in time for this book.

184 **literally feel the pain of others** Singer, T., et al., "Empathy for Pain Involves the Affective but Not Sensory Components of Pain," *Science* 303, no. 5661 (2004): 1157–62.

and, Jackson, P. L., P. Rainville, and J. Decety, "To What Extent Do We Share the Pain of Others? Insight from the Neural Bases of Pain Empathy," *Pain* 125, no. 1–2 (2006): 5–9.

184 **show increased signs of suffering themselves** Langford, D., et al., "Social Modulation of Pain as Evidence for Empathy in Mice," *Science* 312, no. 5782 (2006): 1967–70.

184 **often hug or kiss someone who was bullied** see De Waal, "Putting the Altruism."

and, Palagi, E., and G. Cordoni, "Postconflict Third-Party Affiliation in *Canis lupus*: Do Wolves Share Similarities with the Great Apes?" *Animal Behaviour* 78, no. 4 (2009): 979–86.

185 **tension in the group and preventing future fights** see Aureli, *Natural Conflict Resolution.*

and, Koski, S. E., and E. H. M. Sterck, "Triadic Postconflict Affiliation in Captive Chimpanzees: Does Consolation Console?" *Animal Behaviour* 73, no. 1 (2007): 133–42.

185 **both wolves and dogs display consolation behaviour** see Palagi and Cordoni, "Postconflict Third-Party."

and, Cools, A. K. A., A. J.-M. Van Hout, and M. H. J. Nelissen, "Canine Reconciliation and Third-Party-Initiated Postconflict Affiliation: Do Peacemaking Social Mechanisms in Dogs Rival Those of Higher Primates?" *Ethology* 114, no. 1 (2008): 53–63.

186 **different sex from their primary caretaker** Nagasawa, M., et al., "Dogs Can Discriminate Human Smiling Faces from Blank Expressions," *Animal Cognition* 14, no. 4 (2011): 525–33.

186 **recognize the emotions of others, do not contagiously yawn** Platek, S. M., et al., "Contagious Yawning: The Role of Self-Awareness and Mental State Attribution," *Cognitive Brain Research* 17, no. 2 (2003): 223–27.

and, Senju, A., et al., Absence of Contagious Yawning in Children with Autism Spectrum Disorder. *Biology Letters* 3, no. 6 (2007): 706–8.

186 **even feel similarly through contagion** Joly-Mascheroni, R. M., A. Senju, and A. J. Shepherd, "Dogs Catch Human Yawns," *Biology Letters* 4, no. 5 (2008): 446–48.

186 **indicate that dogs can feel our pain** Three attempts have been made to replicate the contagious yawning phenomenon. Two did not replicate the original finding. One study found no contagious yawning in fifteen dogs but presented the yawning stimuli in video format only (see Harr, A. L., V. R. Gilbert, and K. A. Phillips, "Do Dogs [*Canis Familiaris*] Show Contagious Yawning?" *Animal Cognition* 12, no. 6 [2009]: 833–37). However, another group of researchers was also unable to demonstrate contagious yawning even though they used a human demonstrator (see O'Hara, S. J., and A. V. Reeve, "A Test of the Yawning Contagion and Emotional Connectedness Hypothesis in Dogs, *Canis familiaris*," *Animal Behaviour* 81, no. 1 [2010]: 335–40). Moreover, not everyone agrees that contagious yawning should be interpreted as a measure of empathy (see Yoon, J. M. D., and C. Tennie, "Contagious Yawning: A Reflection of Empathy, Mimicry, or Contagion?"

Animal Behaviour 79, no. 5 [2010], e1–e3). The most recent replication played the sound of a human yawning and found that dogs contagiously yawned in response to a familiar person but not someone they did not know. This result is consistent with an empathy model of contagious yawning in dogs (see Silva, K., J. Bessa, and L. de Sousa, "Auditory Contagious Yawning in Domestic Dogs [*Canis familiaris*]: First Evidence for Social Modulation." *Animal Cognition* 15, no. 4 [2012]: 721–24). Clearly, this will be an exciting area for future research, given the current uncertainty in the literature. For now, what we can be certain of is that dogs react to our emotional responses, but we do not know if they do this because they have an empathic reaction to others.

Chapter 9: Best in Breed (pages 191–217)

192   **recognized by kennel clubs all over the world** Galibert, F., et al., "Toward Understanding Dog Evolutionary and Domestication History," *Biological Psychology* 74, no. 2 (2011): 263–85.

192   **intimidating dog was a mastiff** Ritvo, H., "Pride and Pedigree: The Evolution of the Victorian Dog Fancy," *Victorian Studies* 29, no. 2 (1986): 227–53.

192   **made the meat more tender** see Gregory and Grandin, *Animal Welfare and Meat Science.*

193   **a protruding mandible and wide, flared nostrils** see Ritvo, "Pride and Pedigree."

193   **easiest way to broadcast this was by the dog's appearance** see Ibid.

193   **the same year Darwin published *On the Origin of Species*** Darwin, C., *On the Origin of Species by Means of Natural Selection, or The Preservation of Favoured Races in the Struggle for Life* (New York: D. Appleton, 1860).

193   **which translates roughly to £75,000 today** Available from www .measuringworth.com/ppoweruk.

195   **0.04 percent of the dog genome has evolved** see Wayne and Ostrander, "Lessons Learned."

195   **dogs descended from wolves** Nobel Prize winner Konrad Lorenz was for a time convinced dogs descended from jackals (see Lorenz, K., *Man Meets Dog*, trans. M. K. Wilson [London: Methuen, 1954], and Darwin himself thought some breeds – but not all – descended from wolves (see Darwin, *The Variation of Animals and Plants under Domestication*).

195   **only two major groups of breeds** see Parker, H. G., et al., "Genetic Structure," *Science* 304, no. 5674 (2004): 1160–64.

195   **places due to self-domestication** This result might support the idea that dogs evolved multiple times from multiple populations of wolves. Supporting this hypothesis is the recent finding of a dog-like skull in a Siberian cave dated to 33,000 years ago (Ovodov, N. D., et al., "A 33,000-Year-Old Incipient Dog from the Altai Mountains of Siberia: Evidence of the Earliest Domestication Disrupted by the Last Glacial Maximum," *PLoS ONE* 6, no. 7 [2011], e22821), a date that by far precedes the next most ancient evidence of dog domestication. These authors suggest these proto-dogs represent an extinct

line of wolf-like dogs that was undergoing self-domestication. They also argue the skulls support the idea that dogs evolved multiple times in multiple places from wild wolves (as suggested by Coppinger and Coppinger, *Dogs: A New Understanding*).

195   **dog most closely related to wolves, the basenji** see vonHoldt, "Genome-wide SNP."

195   **breeds that live within the range of modern wolves** see Ibid.

195   **dogs of 'European origin'** There would be many more groups of dogs to consider, but as Europeans colonized the rest of the world, they took their dogs with them and exterminated the genetically distinctive dogs they found in the countries they colonized. For example, few genes from Native American dogs are found in modern dogs in the US. Europeans must have actively prevented interbreeding to keep their dogs as 'pure' lines. Apparently even dogs can suffer from human discrimination (see Castroviejo-Fisher, S., et al., "Vanishing Native American Dog Lineages," *BMC Evolutionary Biology* 11, no. 1 [2011]: 73).

195   **European origin are so small they are barely detectable.** see Parker, et al., "Genetic Structure."

196   **types in dogs are regulated by just three of them** Cadieu, E., et al., "Coat Variation in the Domestic Dog Is Governed by Variants in Three Genes," *Science* 326, no. 5949 (2009): 150–53.

196   **type dogs like the Newfoundlands and Rottweilers** Parker, et al., "Genetic Structure."

196   **the mountain group, along with St Bernards** Parker, H. G., et al., "Breed Relationships Facilitate Fine-Mapping Studies: A 7.8-kb Deletion Cosegregates with Collie Eye Anomaly Across Multiple Dog Breeds," *Genome Research* 17, no. 11 (2007): 1562–71.

196   **Dobermann pinschers and Portuguese water dogs** see vonHoldt, "Genome-wide SNP."

197   **a DNA code like this** You might remember from school biology lessons that the double-stranded DNA that is in all our cells zips and unzips like a zipper. The teeth of the zipper have to fit together, and there are four kinds of zipper heads called nucleotides. The nucleotide adenine always zips together with thymine, while cytosine always zips together with guanine. The letters *GGT* stand for a string of nucleotides on one side of the DNA zipper – in this case guanine, guanine, and thymine. If another individual or species has a different combination of these nucleotides in the same position in their genome, you know that their DNA code is different.

198   **apartment blocks whose owners do not pick up** Glass, Ira, "Witness for the Poo-secution." *This American Life*, November 19, 2010, www .thisamericanlife.org/radio-archives/episode/420/neighborhood -watch?act=3.

199   **and wide behavioural differences** Scott and Fuller, *Genetics and the Social Behavior of the Dog*.

200   **'voice and forced the subject's head from side to side'** see Ibid.

201   **on the lead but howled and wailed** see Ibid.

201   like mazes and detour tests see Ibid.

201   'against accepting the idea of a breed stereotype' see Ibid.

201   'physically and behaviorally excellent animals' see Ibid.

201   dangers of crossbreeding races in humans George, W. C., *The Biology of the Race Problem* http://www.thechristianidentityforum.net/downloads/Biology-Problem.pdf (1962).

202   they chase their tails for hours a day Moon-Fanelli, A., *Canine Compulsive Behavior: An Overview and Phenotypic Description of Tail Chasing in Bull Terriers* http://btca.com/cms_btca/images/documents/updates/canine_compulsive_behavior_fanelli.pdf (1999).

202   to gain widespread acceptance until the 1980s Wiggins, J. S., *The Five-Factor Model of Personality: Theoretical Perspectives* (New York: Guilford Press, 1996).

      and, John, O. P., and S. Srivastava, "The Big-Five Trait Taxonomy: History, Measurement, and Theoretical Perspectives." In *Handbook of Personality; Theory and Research*, 2nd ed., eds. L. A. Pervin and O. P. John (New York: Guilford Press, 1999), 102–38.

202   This amounted to some 18,000 words McCrae, R. R., and O. P. John, "An Introduction to the Five-Factor Model and Its Applications," *Journal of Personality* 60, no. 2 (1992): 175–215.

203   while high neuroticism tends to be a health risk see John and Srivastava, "The Big-Five."

203   tests have become popular in the last decade Kubinyi, E., B. Turcsán, and Á. Miklósi, "Dog and Owner Demographic Characteristics and Dog Personality Trait Associations," *Behavioural Processes* 81, no. 3 (2009): 392–401.

206   bolder dogs are more exploratory Svartberg, K., and B. Forkman, "Personality Traits in the Domestic Dog (*Canis familiaris*)," *Applied Animal Behaviour Science* 79, no. 2 (2002): 133–55.

206   such as obedience, searching, and tracking Svartberg, K., "Shyness–Boldness Predicts Performance in Working Dogs," *Applied Animal Behaviour Science* 79, no. 2 (2002): 157–74.

206   continuum is one that is conserved over time Kagan, J., J. S. Reznick, and N. Snidman, "Biological Bases of Childhood Shyness," *Science* 240, no. 4849 (1988): 167–71.

207   bolder are more successful in killing prey Fox, M. W., "Socio-ecological Implications of Individual Differences in Wolf Litters: A Developmental and Evolutionary Perspective," *Behaviour* 41, nos. 3–4 (1972): 298–313.

207   does your dog learn quickly see Kubinyi, Turcsán, and Miklósi, "Dog and Owner."

      and, Turcsán, B., E. Kubinyi, and Á. Miklósi, "Trainability and Boldness Traits Differ Between Dog Breed Clusters Based on Conventional Breed Categories and Genetic Relatedness," *Applied Animal Behaviour Science* 132, no. 1–2 (2011): 61–70.

208   required reconstructive surgery for dog bites Bini, J. K., et al., "Mortality, Mauling, and Maiming by Vicious Dogs," *Annals of Surgery* 253, no. 4 (2011): 791–97.

209 **bitten by the age of twelve** Hussain, S. G., "Attacking the Dog-Bite Epidemic: Why Breed-Specific Legislation Won't Solve the Dangerous-Dog Dilemma," *Fordham Law Review* 74, no. 5 (2005): 2847–87.

209 **suffer from post-traumatic stress disorder** Peters, V., et al., "Posttraumatic Stress Disorder After Dog Bites in Children," *Journal of Pediatrics* 144, no. 1 (2004): 121–22.

209 **cost US insurance companies $345 million** see Hussain, "Attacking the Dog." and, Bini, et al., "Mortality, Mauling, and Maiming."

209 **cost the NHS £3.3 million** UK Department for Environment, Food and Rural Affairs, "Dangerous Dogs," October 26, 2012, www.defra.org.uk/wildlife-pets/pets/dangerous.

209 **the Dangerous Dogs Act of 1991** Ibid.

209 **the general population and with the public** see Hussain, "Attacking the Dog."

210 **pit bulls or any other dog** see Bini, et al., "Mortality, Mauling, and Maiming."

210 **responsible for 59 percent of these deaths** *US Dog Bite Fatalities January 2006 to December 2008,* April 22, 2009, www.dogsbite.org.

210 **three pitbulls are shot and killed every day because of aggression** see Bini, et al., "Mortality, Mauling, and Maiming."

210 **between 1979 and 1988 involved pit bull types** Overall, K. L., and M. Love, "Dog Bites to Humans – Demography, Epidemiology, Injury, and Risk," *Journal of the American Veterinary Medical Association* 218, no. 12 (2001): 1923–34.

210 **Occasionally, there is such a horrific attack** One horrific case that led to calls to ban pit bulls occurred when a baby was mauled by two pit bulls in April 2009 in San Antonio, Texas. A grandmother was looking after her eleven-month-old grandson, Iziah, and had left him in the bedroom to warm a bottle of milk in the kitchen. When she returned, she found her two pit bulls attacking the baby. She tried to pull the dogs off, but they would not let go. She ran into the kitchen to get a knife and began stabbing her dogs. The dogs turned on her, releasing the boy. When emergency medical technicians arrived, there was blood everywhere, and they could not get past the dogs to reach the baby. Police arrived minutes later and had to shoot the dogs to get to the house (see Bini, et al., "Mortality, Mauling, and Maiming"). The baby was taken to the hospital. Parts of his scalp were bitten to the bone, and there was a deep puncture wound near his neck, as well as bites from his head to the buttocks. The baby died during treatment at the hospital. The grandmother was also hospitalized with serious injuries. She faced criminal charges but died before the case could come to trial.

210 **muzzled, or banned altogether** www.understand-a-bull.com/BSL/Locations/USLocations.htm.

211 **'insurance coverage on the animal'** http://sconet.state.oh.us/rod/docs/pdf/0/2007/2007-ohio-3724.pdf.

212 **even though they were pictures of the same dog** McNicholas, J., and G. M. Collis, "Dogs as Catalysts for Social Interactions: Robustness of the Effect," *British Journal of Psychology* 91, no. 1 (2000): 61–70.

212 'a notable proportion' of bites Monroy, A., et al., "Head and Neck Dog Bites in Children," *Otolaryngology – Head and Neck Surgery* 140, no. 3 (2009): 354–57.

212 top biting culprit was German shepherds Brogan, T. V., et al., "Severe Dog Bites in Children," *Pediatrics* 96, no. 5 (1995): 947–50.

212 English springer spaniels also ranked high Reisner, I. R., F. S. Shofer, and M. L. Nance, "Behavioral Assessment of Child-Directed Canine Aggression," *Injury Prevention* 13, no. 5 (2007): 348–51.

213 dog towards strangers and other dogs was the dachshund Duffy, D. L., Y. Hsu, and J. A. Serpell, "Breed Differences in Canine Aggression," *Applied Animal Behaviour Science* 114, nos. 3–4 (2008): 441–60.

213 not even responsible for most of the injuries see Hussain, "Attacking the Dog."

213 to get killed by a car www-fars.nhtsa.dot.gov/Main/index.aspx.
   and, www.bts.gov/publications/national_transportation_statistics/html/table_01_11.html.

213 to get struck by lightning www.weather.gov/om/lightning/medical.htm.

213 bites happen to children under the age of ten see Overall and Love, "Dog Bites to Humans."

213 children bitten are boys, and 87 percent are white see Brogan, et al., "Severe Dog Bites."

213 contact with the dog's food or possessions see Reisner, Shofer, and Nance, "Behavioral Assessment."

213 average length of the wound being 7.5 centimetres see Monroy, et al., "Head and Neck Dog."

213 bite children have never before bitten a child see Reisner, Shofer, and Nance, "Behavioral Assessment."

213 involved in a serious dog-related injury see Brogan, et al., "Severe Dog Bites."
   and, Overall and Love, "Dog Bites to Humans."

214 retrievers, and poodles are among the cleverest dogs Helton, W. S., "Does Perceived Trainability of Dog (*Canis lupus familiaris*) Breeds Reflect Differences in Learning or Differences in Physical Ability?" *Behavioural Processes* 83, no. 3 (2010): 315–23.

214 a barrier based on a human demonstration Pongrácz, P., et al., "The Pet Dogs Ability for Learning from a Human Demonstrator in a Detour Task Is Independent from the Breed and Age," *Applied Animal Behaviour Science* 90, nos. 3–4 (2005): 309–23.

214 breed differences in the ability of dogs to follow a human pointing gesture Dorey, N. R., M. A. R. Udell, and C. D. L. Wynne, "Breed Differences in Dogs Sensitivity to Human Points: A Meta-Analysis," *Behavioural Processes* 81, no. 3 (2009): 409–15.

214 quite skilled at reading human gestures see Hare, et al., "The Domestication Hypothesis."
   and, Smith and Litchfield, "Dingoes (*Canis dingo*)."
   and, Wobber, V., et al., "Breed Differences in Domestic."

214 the working dogs were better at it see Ibid.

215 **Mariana Bentosela of Buenos Aires University in Argentina** Jakovcevic, A., et al., "Breed Differences in Dogs' (*Canis familiaris*) Gaze to the Human Face," *Behavioural Processes* 84, no. 2 (2010): 602–7.

215 **rearing experiences that may explain the results** see Wobber, et al., "Breed Differences in Domestic."

215 **Moreover, William Helton from the University of Canterbury** Helton, W. S., and N. D. Helton, "Physical Size Matters in the Domestic Dog's (*Canis lupus familiaris*) Ability to Use Human Pointing Cues," *Behavioural Processes* 85, no. 1 (2010): 77–79.

215 **observed in humans than dogs with longer skulls** see Míklósi, *Dog Behaviour, Evolution, and Cognition.*

    and, Helton, W. S., "Cephalic Index and Perceived Dog Trainability," *Behavioural Processes* 82, no. 3 (2009): 355–58.

216 **mean that size is the only reason they vary** see Helton and Helton, "Physical Size."

217 **ranking breeds such as Chihuahuas and shih tzus** see Helton, "Does Perceived Trainability of Dog."

217 **like greyhounds, or broad skulls like bulldogs** see Helton, "Cephalic Index."

Chapter 10: Teaching Genius (pages 219–51)

222 **originated more than two thousand years ago** see vonHoldt, "Genome-wide SNP."

222 **he just was not obeying them** I had been suspicious for a while because Milo was excellent at using my pointing gestures to find hidden food, even though he did not seem to understand my verbal and gestural commands to sit and stay.

223 **As a result, cognitive skills that were there all the time finally** I was probably very lucky with Milo. The research available at the time I had Milo neutered did not suggest a high probability of major behaviour changes as a result of neutering an adult male dog. Some studies suggested neutering male dogs helped decrease obsessive sniffing, marking, and a tendency to stray by reducing circulating androgens such as testosterone that are produced by the testes. Neutering an adult dog does not always result in significant change in behaviour. But it is a good idea to have your dog spayed or neutered anyway!

224 **in the US from 1913 to 1960** Baars, B. J., *The Cognitive Revolution in Psychology* (New York: Guilford Press, 1986).

224 **everyone reviewing your paper was a behaviourist** As George Miller, one of the leaders of the cognitive revolution, said, 'I was educated to study behavior, and I learned to translate my ideas into the new jargon of behaviorism. As I was most interested in speech and hearing, the translation sometimes became tricky. But one's reputation as a scientist could depend on how well the trick was played' (see Miller, G. A., "The Cognitive Revolution: A Historical Perspective," *Trends in Cognitive Sciences* 7, no. 3 [2003]: 141–44).

225 **'find herself hurled to earth from her cloud-castles'** Freud, S., "The Passing of the Oedipus Complex," *International Journal of Psycho-Analysis* 5 (1924): 419–24.

225 **something much more elementary** One of the founders of behaviourism, J.B. Watson, said, 'The time seems to have come when psychology must discard all reference to consciousness; when it need no longer delude itself into thinking that it is making mental states the object of observation'.

225 **'from the unicellular organisms to man'** Watson, "Psychology as the Behaviorist Views It."

225 **Bill Gates, Al Gore, and Einstein** Mooney, C. and S. Kirshenbaum, *Unscientific America: How Scientific Illiteracy Threatens Our Future* (New York: Basic Books, 2009).

225 **In 1975, the best-known scientist in the US was Skinner** O'Donohue, W. T. and K. E. Ferguson, *The Psychology of B. F. Skinner* (Thousand Oaks, Calif.: Sage, 2001).

225 **was on the *New York Times* bestseller list for twenty-six weeks** Rutherford, A., *Beyond the Box: B. F. Skinner's Technology of Behavior from Laboratory to Life, 1950s–1970s* (Toronto: University of Toronto Press, 2009).

226 **'the Hitler of late twentieth century science itself'** Bjork, D. W., *B. F. Skinner: A Life* (New York: Basic Books, 1993).

226 **'visibly insane', and a 'fatuous opinionated ass'** see Rutherford, *Beyond the Box*.

226 **He was considered conceited even at university** see Bjork, *B. F. Skinner*.

226 **Skinner described the entire event with remarkable detachment** see Ibid.

226 **'be greater through his work than through his genes'** Hothersall, D., *History of Psychology* (New York: Random House, 1984).

226 **Skinner discovered behaviourism** see Bjork, *B. F. Skinner*.

226 **salivating at the sight of their keeper before there was any food in sight** see Hunt, *The Story of Psychology*

227 **animal's behaviour becomes the stimulus** Shettleworth, *Cognition, Evolution, and Behavior*.

227 **which became known as 'rat basketball'** see Hothersall, *History of Psychology*.

227 **guide the bombs towards their target** see O'Donohue, *The Psychology of B. F. Skinner*.

228 **and it certainly did not have any place in psychology** see Ibid.

228 **knowing if they were telling the truth** see Ibid.

228 **trailer that he could tow to schools around Seattle** see Rutherford, *Beyond the Box*.

229 **physically and mentally challenged in skills worksops, and juveniles in delinquency centres** see Ibid.

229 **token economy programmes in twenty hospitals, involving more than nine hundred patients** see Ibid.

229 **this was completely ignored** There was only one study on the reaction of patients to the token economy. Biklen remarked that only one or two patients seemed to respond positively to the programme. The rest wavered

between resentment and withdrawal (Biklen, D. P., "Behavior Modification in a State Mental Hospital," *American Journal of Orthopsychiatry* 46, no. 1 [1976]: 53–61). Although presumably the outcome of the programme was the desired modification of the patients' behaviour, Biklen both observed and experienced significant anger towards himself and the students carrying out the programme.

'They can go shove the whole thing as far as I'm concerned,' one patient told him. 'Go to hell with your stars,' another woman said to a student who suggested she accumulate stars in exchange for cigarettes.

Biklin found that once the patients discovered he was not part of the programme and had no power to dispense stars for good behaviour, they became friendly towards him. They sat down and talked about life before the institution, their frustrations, and their desires of being released (see Ibid.).

The behaviour modification programme was supposed to improve the lives of the patients and staff, but ironically talking about thoughts and feelings gave them more satisfaction.

229   **Skinner did not find inflicting pain on animals productive** see O'Donohue, *The Psychology of B. F. Skinner.*

229   **He kept his rats at about 80 percent of their normal weight** see Ibid.

229   **the mentally impaired, dying, and prisoners** see Rutherford, *Beyond the Box.*

230   **weight gain, smoking, speech problems, autism** see Hothersall, *History of Psychology.*

230   **Behaviourism depended on four things** Chomsky, N., "A Review of B. F. Skinner's *Verbal Behavior*," *Language* 35, no. 1 (1959): 26–58.
       and, Watson, "Psychology as the Behaviorist Views It."

231   **having another dog around did help** Dreschel, N. A. and D. A. Granger, "Physiological and Behavioral Reactivity to Stress in Thunderstorm-Phobic Dogs and Their Caregivers," *Applied Animal Behaviour Science* 95, nos. 3–4 (2005): 153–68.

232   **fireworks, bees, and high-altitude aircraft** Rogerson, J., "Canine Fears and Phobias: A Regime for Treatment Without Recourse to Drugs," *Applied Animal Behaviour Science* 52, nos. 3–4 (1997): 291–97.

232   **that Skinner's reputation never quite recovered** see Baars, "The Cognitive Revolution."
       and, Chomsky, "A Review."

233   **Skinner and the school of behaviourism** see Tomasello and Call, "Primate Cognition."
       and, Shettleworth, *Cognition, Evolution, and Behavior.*

233   **instruction how to use their culture's language** Tomasello, M., *Constructing a Language: A Usage-Based Theory of Language Acquisition* (Cambridge, Mass.: Harvard University Press, 2005).

234   **sit for a new person in a different location** Thorn, J. M., et al., "Conditioning Shelter Dogs to Sit," *Journal of Applied Animal Welfare Science* 9, no. 1 (2006): 25–39.

237 **dominance towards their human partners** Rooney, N. J., and J. W. S. Bradshaw, "An Experimental Study of the Effects of Play Upon the Dog-Human Relationship," *Applied Animal Behaviour Science* 75, no. 2 (2002): 161–76.

237 **unlikely to enjoy reading for its o-wn sake** Deci, E. L., R. Koestner, and R. M. Ryan, "A Meta-Analytic Review of Experiments Examining the Effects of Extrinsic Rewards on Intrinsic Motivation," *Psychological Bulletin* 125, no. 6 (1999): 627–68.

and, Warneken, F., and M. Tomasello, "Extrinsic Rewards Undermine Altruistic Tendencies in 20-Month-Olds." *Developmental Psychology* 44, no. 6 (2008): 1785–88.

238 **treats if you run out of the good stuff** Bentosela, M., et al., "Incentive contrast in domestic dogs (*Canis familiaris*)," *Journal of Comparative Psychology* 123, no. 2 (2009): 125–30.

238 **training session a week learned in fewer sessions** Meyer, I., and J. Ladewig, "The Relationship Between Number of Training Sessions Per Week and Learning in Dogs," *Applied Animal Behaviour Science* 111, nos. 3–4 (2008): 311–20.

238 **dogs receiving the most training – every day for three sessions** Demant, H., et al., "The Effect of Frequency and Duration of Training Sessions on Acquisition and Long-Term Memory in Dogs," *Applied Animal Behaviour Science* 133, nos. 3–4 (2011): 228–34.

238 **most popular training techniques today** Pryor, K., *Don't Shoot the Dog!* (California: Ringpress Books, 1999).

and, Pryor, K., *Getting Started: Clicker Training for Dogs* (Waltham, Mass.: Sunshine, 2002).

239 **no scientific evidence to support the theory that clicker training facilitates faster learning in dogs** Smith, S. M., and E. S. Davis, "Clicker Increases Resistance to Extinction but Does Not Decrease Training Time of a Simple Operant Task in Domestic Dogs (*Canis familiaris*)," *Applied Animal Behaviour Science* 110, no. 3 (2008): 318–29.

240 **detour around a physical barrier than dogs** see Frank, et al., "Motivation and Insight in Wolf."

240 **the results are not very good** Moreover, there is currently little, if any, published evidence for major breed differences in the ability to learn or in the rate of learning. The only relevant data set is decades old and found inconsistent breed differences depending on the learning tasks being used (see Scott and Fuller, *Genetics and the Social Behavior*). One major factor has been seen consistently to play an important role across different learning studies: Younger dogs tend to outperform older dogs in terms of the speed of their learning and their memory for where food is hidden (for example, see Marshall-Pescini, S., et al., "Does Training Make You Smarter? The Effects of Training on Dogs' Performance [*Canis familiaris*] in a Problem Solving Task," *Behavioural Processes* 78, no. 3 [2008]: 449–54; and, Milgram, N., et al., "Learning Ability in Aged Beagle Dogs Is Preserved by Behavioral Enrichment and Dietary Fortification: A Two-Year Longitudinal Study,"

Neurobiology of Aging 26, no. 1 [2005]: 77–90). Otherwise we have little evidence that different types of dogs consistently differ in their ability to learn. These age effects should not be interpreted to mean that older dogs cannot learn new tricks. It only means that younger dogs are generally quicker on the uptake. Given my experience with Milo, it is also easy to imagine that future research might find consistent differences in how different breeds learn. The absence of evidence for differences in this case should be interpreted cautiously, since so few studies have been conducted.

240 **using human gestures as a result of either training** see McKinley and Sambrook, "Use of Human-Given Cues."

and, Wobber and Hare, "Testing the Social Dog Hypothesis."

240 **Even rescue dogs and breeds not intentionally bred** see Ibid.

and, Smith, B. P., and C. A. Litchfield, "How Well Do Dingoes, *Canis dingo*, Perform on the Detour Task?" *Animal Behaviour* 80, no. 1 (2010): 155–62.

and, Hare, et al., "The Domestication Hypothesis."

but, Udell, Dorey, and Wynne, "The Performance of Stray Dogs."

241 **attention to them while giving the gesture** see Téglás, et al., "Dogs' Gaze Following Is Tuned."

and, see Kaminski, Schulz, and Tomasello, "How Dogs Know When."

241 **dogs are less likely to use your pointing gesture** see Ibid.

241 **threatening gestures to prevent them from going somewhere** Pettersson, H., et al., "Understanding of Human Communicative Motives in Domestic Dogs," *Applied Animal Behaviour Science* 133, no. 3 (2011): 235–45.

241 **high-pitched voice as opposed to a low-pitched** Scheider, L., et al., "Domestic Dogs Use Contextual Information and Tone of Voice When Following a Human Pointing Gesture," *PLoS ONE* 6, no. 7 (2011): e21676.

241 **pay attention and learn from the demonstration** see Pongrácz, et al., "Social Learning in Dogs."

and, Pongrácz, et al., "Verbal Attention Getting."

241 **command in a known location** Braem, M. D., and D. S. Mills, "Factors Affecting Response of Dogs to Obedience Instruction: A Field and Experimental Study," *Applied Animal Behaviour Science* 125, no. 1 (2010): 47–55.

242 **trained dogs persisted until they found a solution** see Miklósi, "A Simple Reason for a Big."

and, Marshall-Pescini, et al., "Does Training Make You Smarter?"

242 **highest training scores than other puppies** Slabbert, J., and O. A. E. Rasa, "Observational Learning of an Acquired Maternal Behaviour Pattern by working Dog Pups: An Alternative Training Method?" *Applied Animal Behaviour Science* 53, no. 4 (1997): 309–16.

242 **they see someone else solve the problem first** see Miller, Rayburn-Reeves, and Zentall, "Imitation and Emulation."

and, Pongrácz, et al., "Social Learning in Dogs."

and, Frank, et al., "Motivation and Insight in Wolf."

243 **fetch a ball that a human can see** see Hare, Call, and Tomasello, "Communication of Food."

and, Kaminski, Schulz, and Tomasello, "How Dogs Know."

and, Virányi, et al., "Dogs Respond Appropriately."

and, Gácsi, et al., "Are Readers of Our Face."

and, Fukuzawa, Mills, and Cooper, "More Than Just a Word."

and, Schwab and Huber, "Obey or Not Obey?"

243   **Christine Schwab and Ludwig Huber from the University of Vienna** see Schwab and Huber, "Obey or Not Obey?"

and, Virányi, et al., "Dogs Respond Appropriately."

and, Call, et al., "Domestic dogs (*Canis familiaris*)."

246   **their ability to rely on what they can see** Osthaus, Marlow, and Ducat "Minding the Gap."

246   **instead go to where the human pointed** see Szetei, et al., "When Dogs Seem to Lose Their Nose."

246   **their sense of smell might miss their target** Lit, L., J. B. Schweitzer, and A. M. Oberbauer, "Handler Beliefs Affect Scent Detection Dog Outcomes," *Animal Cognition* 14, no. 3 (2011): 387–94.

246   **their right paw and males favouring their left** Wells, D. L., "Lateralised Behaviour in the Domestic Dog, *Canis familiaris*," *Behavioural Processes* 61, nos. 1–2 (2003): 27–35.

and, Poyser, F., C. Caldwell, and M. Cobb, "Dog Paw Preference Shows Lability and Sex Differences," *Behavioural Processes* 73, no. 2 (2006): 216–21.

246   **brain, leading them to turn their head to the left in response** Siniscalchi, M., et al., "Dogs Turn Left to Emotional Stimuli," *Behavioural Brain Research* 208, no. 2 (2010): 516–21.

and, Siniscalchi, M., A. Quaranta, and L. J. Rogers, "Hemispheric Specialization in Dogs for Processing Different Acoustic Stimuli," *PLoS ONE* 3, no. 10 (2008): e3349.

247   **that dogs react to their owner's frustrated behaviour** Horowitz, A., *Inside of a Dog: What Dogs See, Smell, and Know* (New York: Scribner, 2009).

247   **this had was to raise the stress levels of the dogs** Jones, A. C., and R. A. Josephs, "Interspecies Hormonal Interactions Between Man and the Domestic Dog (*Canis familiaris*)," *Hormones and Behavior* 50, no. 3 (2006): 393–400.

247   **the human had or had not seen the object hidden** see Kaminski, "Dogs, *Canis familiaris*."

but, see Topál, J., Á. E. R. Mányik, and Á. Miklósi, "Mindreading in a Dog: An Adaptation of a Primate 'Mental Attribution' Study," *International Journal of Psychology and Psychological Therapy* 6, no. 3 (2006): 365–79.

248   **understand the threat of strangers** see Ortolani, Vernooij, and Coppinger, "Ethiopian Village Dogs."

and, Bonanni, et al., "Free-Ranging Dogs."

248   **owner was crying in pain and calling for help** Macpherson, K., and W. A. Roberts, "Do Dogs (*Canis familiaris*) Seek Help in an Emergency?" *Journal of Comparative Psychology* 120, no. 2 (2006): 113–19.

249   **whether it is by making inferences like Rico and Chaser** see Kaminski, Call, and Fisher, "Word Learning."

and, Pilley and Reid, "Border Collie Comprehends."

249 **paying attention to our communicative gestures** see Miklósi and Soproni, "A Comparative Analysis."

and, Hare, B., and M. Tomasello, "Human-like Social Skills in Dogs?" *Trends in Cognitive Sciences* 9, no. 9 (2005): 439–44.

249 **learning to learn** see Marshall-Pescini, et al., "Does Training Make You Smarter?"

249 **watching someone else do it first** see Miller, Rayburn-Reeves, and Zentall, "Imitation and Emulation."

and, Pongrácz, et al., "Social Learning in Dogs."

and, Crucially, simple conditioned learning, inferential abilities, or social learning mechanisms are not necessarily in competition with one another as explanations for dog behaviour. Evidence for improved performance does not necessarily rule out the possibility of high-level cognitive abilities or the use of inferential skills. Likewise, the presence of inferential skills does not rule out the possibility for further learning through trial and error (see Hare, et al., "The Domestication Hypothesis"; and, Call, J., "Chimpanzee Social Cognition," *Trends in Cognitive Sciences* 5, no. 9 [2001]: 388–93). Evidence for more inflexible learning processes such as operant and classical conditioning would be found if dogs learned one problem but could not generalize this learning to a slightly different situation. If dogs had to slowly learn over dozens or hundreds of trials when exposed to a slight variation on a previously learned game, then it is likely they are using inflexible forms of conditioned learning.

250 **everyone else in the neighbourhood is barking** Cohen, J. A., and M. W. Fox, "Vocalizations in Wild Canids and Possible Effects of Domestication," *Behavioural Processes* 1, no. 1 (1976): 77–92.

250 **one minute on command can reduce barking** Yin, S., et al., "Efficacy of a Remote-Controlled, Positive-Reinforcement, Dog-Training System for Modifying Problem Behaviors Exhibited When People Arrive at the Door," *Applied Animal Behaviour Science* 113, nos. 1–3 (2008): 123–38.

250 **for a period of three weeks** Wells, D. L., "The Effectiveness of a Citronella Spray Collar in Reducing Certain Forms of Barking in Dogs," *Applied Animal Behaviour Science* 73, no. 4 (2001): 299–309.

250 **electronic shock collar in reducing barking** Steiss, J. E., et al., "Evaluation of Plasma Cortisol Levels and Behavior in Dogs Wearing Bark Control Collars," *Applied Animal Behaviour Science* 106, nos. 1–3 (2007): 96–106.

Chapter 11: For the Love of Dog (pages 253–82)

253 **In other places and cultures, it is different** Herzog, H., *Some We Love, Some We Hate, Some We Eat: Why It's So Hard to Think Straight About Animals* (New York: HarperCollins, 2010).

254 **dogs vary historically and around the world** see Ibid.

254 **'North America, and then buried affectionately'** see Morey, *Dogs.*

255 **suggesting dogs were being eaten** see Ibid.

255 **puppy to lick their bottoms clean** see Coppinger and Coppinger, *Dogs.*

255 **there are thousands of stray dogs** Alie, K., et al., "Attitudes Towards Dogs and Other 'Pets' in Roseau, Dominica," *Anthrozoös* 20, no. 2 (2007): 143–54.
and, Davis, B. W., et al., "Preliminary Observations on the Characteristics of the Owned Dog Population in Roseau, Dominica," *Journal of Applied Animal Welfare Science* 10, no. 2 (2007): 141–51.

256 **the Japanese after they were dead** Veldkamp, E., "The Emergence of 'Pets as Family' and the Socio-Historical Development of Pet Funerals in Japan," *Anthrozoös* 22, no. 4 (2009): 333–46.

256 **owners, and clubbed to death on the spot** French, H. W., "A Chinese Outcry: Doesn't a Dog Have Rights?" *The New York Times*, August 10, 2006, www.nytimes.com/2006/08/10/world/asia/10china.html.

257 **contracted rabies were not even given the right drugs** Tang, X., et al., "Pivotal Role of Dogs in Rabies Transmission, China," *Emerging Infectious Diseases* 11, no. 12 (2005): 1970–72.

257 **emperors to students studying for exams** Morgan, C., "Dogs and Horses in Ancient China," *Journal of the Royal Asiatic Society, Hong Kong Branch* 14 (1974): 58–68.

257 **people knelt down as they passed** Collier, V. W. F., *Dogs of China and Japan in Nature and Art* (New York: Frederick A Stokes 1921).

258 **receive was that his dogs could go 'in the book'** see Collier, *Dogs of China and Japan.*

259 **every city to destroy all of the dogs** Kinmond, W., *No Dogs in China: A Report on China Today* (New York: Thomas Nelson and Sons, 1957).

259 **they were banned from the city of Beijing** Wines, M., "Once Banned, Dogs Reflect China's Rise," *The New York Times*, October 24, 2010, www.nytimes.com/2010/10/25/world/asia/25dogs.html.

259 **canine army to conquer western Europe** Li, Y., et al., "The Origin of the Tibetan Mastiff and Species Identification of *Canis* Based on Mitochondrial Cytochrome C Oxidase Subunit I (COI) Gene and COI Barcoding," *animal* 5, no. 12 (2011): 1868–73.

259 **diet includes abalone and sea cucumber** McGraw, S., "$1.5 Million Paid for World's Most Expensive Dog," *Today*, March 17, 2011, http://today.msnbc.msn.com/id/42128943/ns/today-today_pets_and_animals/t/million-paid-worlds-most-expensive-dog.

259 **drink at a 'doggy bar' downtown** Wines, "Once Banned, Dogs Reflect."

260 **around half are euthanized** see Range, et al., "The Effect of Ostensive Cues."

260 **forbidding pets on the property** Salman, M., et al., "Human and Animal Factors Related to Relinquishment of Dogs and Cats in 12 Selected Animal Shelters in the United States," *Journal of Applied Animal Welfare Science* 1, no. 3 (1998): 207-26.

260 **reported that their dog soiled the house** New, J. C., et al., "Moving: Characteristics of Dogs and Cats and those relinquishing them to 12 US Animal Shelters," *Journal of Applied Animal Welfare Science* 2, no. 2 (1999): 83–96.

261 **often suffer from injuries to their paws** Pacelle, W., *The Bond: Our Kinship with Animals, Our Call To Defend Them* (New York: William Morrow, 2011).

261 'Two of the pups are dead' Fumarola, A. J., "With Best Friends Like Us Who Needs Enemies? The Phenomenon of the Puppy Mill, the Failure of Legal Regimes to Manage It, and the Positive Prospects of Animal Rights." *Buffalo Environmental Law Journal* 6, no. 2 (1999) 253.

261 supplier of puppies to the public see Ibid.

262 four million puppies each year see Pacelle, *The Bond*.

262 keep different breeds in stock Summers, K., manager, puppy mills campaign, Humane Society of the US (personal communication, September 6, 2011).

262 to repair through surgery Savino, S. K., "Puppy Lemon Laws: Think Twice Before Buying That Doggy in the Window," *Penn State Law Review* 114, no. 2 (2009): 643–66.

262 web are unlicensed by the USDA Schalke, E., et al., "Clinical Signs Caused by the Use of Electric Training Collars on Dogs in Everyday Life Situations," *Applied Animal Behaviour Science* 105, no. 4 (2007): 369–80.

263 without much interference from the law Kalof, L., and C. Taylor, "The Discourse of Dog Fighting," *Humanity & Society* 31, no. 4 (2007): 319–33.

263 it was not outlawed in all fifty US states until 1976 Gibson, H., "Dog Fighting Detailed Discussion," Animal Legal and Historical Center, Michigan State University College of Law, 2005, http://www.animallaw.info/articles/ddusdogfighting.htm.

263 It has reached epidemic proportions see Kalof and Taylor, "The Discourse."

263 In Birmingham, youth gangs will settle scores BBC News, "Public Dog Fights in Birmingham Concerns City Council," March 16, 2012, www.bbc.co.uk/news/uk-england-birmingham-17383231..

264 easily more than that from an armed robbery see Gibson, "Dog fighting."

263 children conducting their own dogfights see Ibid.

264 'not from the ghetto that make dog fighting happen in Detroit' see Ibid.

265 keeping him warm and saving his life www.purina.ca/about/halloffame/inductee/2000/elmo.aspx. Purina Animal Hall of Fame, 2000.

266 David Tuber from Ohio State University Tuber, D. S., et al., "Behavioral and Glucocorticoid Responses of Adult Domestic Dogs (*Canis familiaris*) to Companionship and Social Separation," *Journal of Comparative Psychology* 110, no. 1 (1996): 103–8.

266 likely to choose the small amount Prato-Previde, E., S. Marshall-Pescini, and P. Valsecchi, "Is Your Choice My Choice? The Owners' Effect on Pet Dogs' (*Canis lupus familiaris*) Performance in a Food Choice Task," *Animal Cognition* 11, no.1 (2008): 167–74.

266 will choose the wrong cup more often Szetei, et al., "When Dogs Seem to Lose Their Nose."

267 infants when they are deprived of their mothers Bretherton, I., "The Origins of Attachment Theory: John Bowlby and Mary Ainsworth," *Developmental Psychology* 28, no. 5 (1992): 759–75.

268 Topál decided to use the Strange Situation test on dogs. Topál, J., et al., "Attachment Behavior in Dogs."

268 Other researchers found that when owners left Prato-Previde, E., et al., "Is the Dog-Human Relationship an Attachment Bond? An Observational

Study Using Ainsworth's Strange Situation," *Behaviour* 140, no. 2 (2003): 225–54.

268 **tried the Strange Situation test with rescue dogs** Gácsi, M., et al., "Attachment Behavior of Adult Dogs (*Canis familiaris*) Living at Rescue Centers: Forming New Bonds," *Journal of Comparative Psychology* 115, no. 4 (2001): 423–31.

269 **influence the city's upcoming mayoral election** Duff-Brown, B., "San Francisco Dog Owners Hope to Sway Mayoral Race," yourlife.usatoday. com/parenting-family/pets/dogs/story/2011-10-04/San-Francisco-dog-owners-hope-to-sway-mayoral-race/50655974/1.

269 **stories of babies and puppies in pain** Angantyr, M., J. Eklund, and E. M. Hansen, "A Comparison of Empathy for Humans and Empathy for Animals," *Anthrozoös* 24, no. 4 (2011): 369–77.

269 **taken a sick day to take care of their dog** Milo's Kitchen Pet Parent Survey, conducted by Kelton Research, April 2011.

and, Youde, Kate, "Pampered Pets UK," *The Independent*, May 6, 2012.

270 **Most men would happily take those odds** Guéguen, N., and S. Ciccotti, "Domestic Dogs as Facilitators in Social Interaction: An Evaluation of Helping and Courtship Behaviors," *Anthrozoös* 21, no. 4 (2008): 339–49.

270 **a young American undergraduate used a black Labrador** see McNicholas and Collis, *Dogs as Catalysts*.

271 **the golden retriever puppy was by far the winner** Wells, D. L., "The Facilitation of Social Interactions by Domestic Dogs," *Anthrozoös* 17, no. 4 (2004): 340–52.

271 **more relaxed, and more approachable** Rossbach, K. A., and J. P. Wilson, "Does a Dog's Presence Make a Person Appear More Likable?: Two Studies," *Anthrozoös* 5, no. 1 (1992): 40–51.

271 **situations besides looking for love** see Pacelle, *The Bond*.

272 **car accidents, drinking problems, and even suicide** Gilbey, A., J. McNicholas, and G. M. Collis, "A Longitudinal Test of the Belief That Companion Animal Ownership Can Help Reduce Loneliness," *Anthrozoös* 20, no. 4 (2007): 345–53.

and, Lynch, J. J., *The Broken Heart: The Medical Consequences of Loneliness* (New York: Basic Books, 1977).

272 **can make people less lonely** Banks, M. R., and W. A. Banks, "The Effects of Animal-Assisted Therapy on Loneliness in an Elderly Population in Long-Term Care Facilities," *Journals of Gerontology Series A: Biological Sciences and Medical Sciences* 57, no. 7 (2002): M428–32.

and, Heath, D. T., and P. C. McKenry, "Potential Benefits of companion Animals for Self-Care Children. Reviews of Research," *Childhood Education* 65, no. 5 (1989): 311–14.

and, Kehoe, M., "Loneliness and the aging Homosexual: Is Pet Therapy an Answer?" *Journal of Homosexuality* 20, nos. 3–4 (1991): 137–42.

and, Mader, B., L. A. Hart, and B. Bergin, "Social Acknowledgments for Children with Disabilities: Effects of Service Dogs," *Child Development* 60, no. 6 (1989): 1529–34.

272 **University in New Zealand and colleagues** see Gilbey, McNicholas, and Collis, "A Longitudinal Test."

272 **their dog than to their fathers or brothers** Kurdek, L. A., "Young Adults' Attachment to Pet Dogs: Findings from Open-Ended Methods," *Anthrozoös* 22, no. 4 (2009): 359–69.

272 **best friend in staving off negative feelings** McConnell, A. R., et al., "Friends with Benefits: On the Positive Consequences of Pet Ownership," *Journal of Personality* 101, no. 6 (2011): 1239–52.

273 **in the company of their dog rather than their best friend** Allen, K. M., et al., "Presence of Human Friends and Pet Dogs as Moderators of Autonomic Responses to Stress in Women," *Journal of Personality and Social Psychology* 61, no. 4 (1991): 582–89.

273 **'I know my partner really loves me'** Beck, L., and E. A. Madresh, "Romantic Partners and Four-Legged Friends: An Extension of Attachment Theory to Relationships with Pets," *Anthrozoös* 21, no. 1 (2008): 43–56.

273 **better at mathematical problems with their pets present** Allen, K., B. E. Shykoff, and J. L. Izzo, "Pet Ownership, but Not ACE Inhibitor Therapy, Blunts Home Blood Pressure Responses to Mental Stress," *Hypertension* 38, no. 4 (2001): 815–20.

273 **another who was cured of a large growth on his neck** Serpell, J. A., "Animal Companions and Human Well-Being: An Historical Exploration of the Value of Human-Animal Relationships." In *Handbook on Animal-Assisted Therapy: Theoretical Foundations and Guidelines for Practice*, 3rd ed., ed. A. H. Fine (New York, Academic Press, 2010), 3–19.

274 **still be alive a year later** Friedmann, E., et al., "Animal Companions and One-Year Survival of Patients After Discharge from a Coronary Care Unit," *Public Health Reports* 95, no. 4 (1980): 307–12.

274 **year after a heart attack than non-dog-owners** Friedmann, E., and S. A. Thomas, "Pet Ownership, Social Support, and One-Year Survival After Acute Myocardial Infarction in the Cardiac Arrhythmia Suppression Trial (CAST)," *American Journal of Cardiology* 76, no. 17 (1995): 1213–17.

274 **as well as a faster recovery** Allen, K., J. Blascovich, and W. B. Mendes, "Cardiovascular Reactivity and the Presence of Pets, Friends, and Spouses: The Truth About Cats and Dogs," *Psychosomatic Medicine* 64, no. 5 (2002): 727–39.

274 **doctor visits than non-owners, even in difficult times** Siegel, J. M., "Stressful Life Events and Use of Physician Services Among the Elderly: The Moderating Role of Pet Ownership," *Journal of Personality and Social Psychology* 58, no. 6 (1990): 1081–86.

275 **'which they share their lives are not included'** Beck, A., and L. Glickman, "Future Research on Pet Facilitated Therapy: A Plea for Comprehension Before Intervention," Technology Assessment Workshop, 1987.

275 **and were more likely to be smokers** Parslow, R., and A. F. Jorm, "Pet ownership and Risk Factors for Cardiovascular Disease: Another Look," *Medical Journal of Australia* 179, no. 9 (2003): 466–68.

275 **call Friedmann's original study 'a canard'** Parker, G., et al., "Survival Following an Acute Coronary Syndrome: A Pet Theory Put to the Test," *Acta Psychiatrica Scandinavica* 121, no. 1 (2010): 65–70.

277 **David would spend time with Ruby** Sockalingam, S., et al., "Use of Animal-Assisted Therapy in the Rehabilitation of an Assault Victim with a

Concurrent Mood Disorder," *Issues in Mental Health Nursing* 29, no. 1 (2008): 73–84.

277 **after the therapy dog visit than the play session** Kaminski, M., T. Pellino, and J. Wish, "Play and Pets: The Physical and Emotional Impact of Child-Life and Pet Therapy on Hospitalized Children," *Children's Health Care* 31, no. 4 (2002): 321–36.

277 **without any drugs, for at least three hours** Braun, C., et al., "Animal-Assisted Therapy as a Pain Relief Intervention for Children," *Complementary Therapies in Clinical Practice* 15, no. 2 (2009): 105–9.

278 **have to be restrained when a dog was present** Hansen, K. M., et al., "Companion Animals Alleviating Distress in Children," *Anthrozoös* 12, no. 3 (1999): 142–48.

278 **aggressive when they interacted with a therapy dog** Filan, S. L., and R. H. Llewellyn-Jones, "Animal-Assisted Therapy for Dementia: A Review of the Literature," *International Psychogeriatrics* 18, no. 4 (2006): 597–612.

and, Barker, S. B., and K. S. Dawson, "The Effects of Animal-Assisted Therapy on Anxiety Ratings of Hospitalized Psychiatric Patients," *Psychiatric Services* 49, no. 6 (1998): 797–801.

and, Johnson, R. A., et al., "Human-Animal Interaction: A Complementary/Alternative Medical (CAM) Intervention for Cancer Patients," *American Behavioral Scientist* 47, no. 1 (2003): 55–69.

279 **Pets are an extra form of social support, not a replacement** see McConnell, et al., "Friends with Benefits."

279 **along nerve fibres to the nervous system** Moberg, K. U., *The Oxytocin Factor: Tapping the Hormone of Calm, Love, and Healing* (Boston: Merloyd Lawrence Books, 2003).

279 **In a study from Japan, people whose dogs gazed at them** Nagasawa, M., et al., "Dog's Gaze at its Owner Increases Owner's Urinary Oxytocin During Social Interaction," *Hormones and Behavior* 55, no. 3 (2009): 434–41.

279 **a room that was empty** Odendaal, J., and R. Meintjes, "Neurophysiological Correlates of Affiliative Behaviour Between Humans and Dogs," *Veterinary Journal* 165, no. 3 (2003): 296–301.

280 **Suzanne Miller and colleagues** Miller, S. C., et al., "An Examination of Changes in Oxytocin Levels in Men and Women Before and After Interaction with a Bonded Dog," *Anthrozoös* 22, no. 1 (2009): 31–42.

280 **its effects on women are more pronounced** see Moberg, *The Oxytocin Factor.*

280 **researchers analysed the testosterone and cortisol** see Jones and Joseph, "Interspecies Hormonal Interactions."

281 **barking and maintaining eye contact** Wells, D. L., and P. G. Hepper, "Male and Female Dogs Respond Differently to Men and Women," *Applied Animal Behaviour Science* 61, no. 4 (1999): 341–49.

281 **In an animal rescue home in Dayton, Ohio** Hennessy, M. B., et al., "Plasma Cortisol Levels of Dogs at a County Animal Shelter," *Physiology & Behavior* 62, no. 3 (1997): 485–90.

282 **the men underwent 'petting training'** Hennessy, M. B., et al., "Influence of Male and Female Petters on Plasma Cortisol and Behaviour: Can Human Interaction Reduce the Stress of Dogs in a Public Animal Shelter?" *Applied Animal Behaviour Science* 61, no. 1 (1998): 63–77.

INDEX

Note:
Page numbers in *italics* refer to illustrations.
Page numbers followed by [t] refer to a table.

355